# LINEAR AC[CELERATORS]
# FOR RADIATI[ON]

## Second Edition

*Other books in the series*

**The Physics and Radiobiology of Fast Neutron Beams**
D K Bewley

**Biomedical Magnetic Resonance Technology**
C-N Chen and D I Hoult

**Rehabilitation Engineering Applied to Mobility and Manipulation**
R A Cooper

**Health Effects of Exposure to Low-Level Ionizing Radiation**
W R Hendee and F M Edwards

**Introductory Medical Statistics**
R F Mould

**Radiation Protection in Hospitals**
R F Mould

**RPL Dosimetry—Radiophotoluminescence in Health Physics**
J A Perry

**Physics of Heart and Circulation**
J Strackee and N Westerhof

**The Physics of Medical Imaging**
S Webb

**The Physics of Three-Dimensional Radiation Therapy: Conformal Radiotherapy, Radiosurgery and Treatment Planning**
S Webb

**The Physics of Conformal Radiotherapy: Advances in Technology**
S Webb

**The Design of Pulse Oximeters**
J G Webster

*Other Medical Physics handbooks*

**Nuclear Particles in Cancer Treatment**
J F Fowler

**Physical Aspects of Brachytherapy**
T J Godden

**Radiation Protection**
R L Kathren

**Physics of Electron Beam Therapy**
S C Klevenhagen

**Thermoluminescence Dosimetry**
A F McKinlay

**Non-ionising Radiation**
H Moseley

Medical Science Series

# LINEAR ACCELERATORS FOR RADIATION THERAPY
## Second Edition

David Greene

Former Assistant Director

Peter C Williams

Deputy Director
North Western Medical Physics Department
The Christie Hospital
Manchester, UK

Institute of Physics Publishing
Bristol and Philadelphia

© IOP Publishing Ltd 1985, 1997

All rights reserved. No part of this publication may be reproduced, stored in a retrieval system or transmitted in any form or by any means, electronic, mechanical, photocopying, recording or otherwise, without the prior permission of the publisher. Multiple copying is permitted in accordance with the terms of licences issued by the Copyright Licensing Agency under the terms of its agreement with the Committee of Vice-Chancellors and Principals.

*British Library Cataloguing-in-Publication Data*

A catalogue record for this book is available from the British Library.

ISBN 0 7503 0402 2 (hbk)
ISBN 0 7503 0476 6 (pbk)

*Library of Congress Cataloging-in-Publication Data are available*

First published 1985

The author has attempted to trace the copyright holder of all the figures and tables reproduced in this publication and apologizes to copyright holders if permission to publish in this form has not been obtained.

Series Editors:
 **R F Mould**, Croydon, UK
 **C G Orton**, Karamanos Cancer Institute, Detroit, USA
 **J A E Spaan**, University of Amsterdam, The Netherlands
 **J G Webster**, University of Wisconsin-Madison, USA

Published by Institute of Physics Publishing, wholly owned by The Institute of Physics, London

Institute of Physics Publishing, Dirac House, Temple Back, Bristol BS1 6BE, UK

US Editorial Office: Institute of Physics Publishing, The Public Ledger Building, Suite 1035, 150 South Independence Mall West, Philadelphia, PA 19106, USA

Printed in Great Britain by J W Arrowsmith Ltd, Bristol

The Medical Science Series is the official book series of the International Federation for Medical and Biological Engineering (IFMBE) and the International Organization for Medical Physics (IOMP).

## IFMBE

The IFMBE was established in 1959 to provide medical and biological engineering with an international presence. The Federation has a long history of encouraging and promoting international cooperation and collaboration in the use of technology for improving the health and life quality of man.

The IFMBE is an organization that is mostly an affiliation of national societies. Transnational organizations can also obtain membership. At present there are 42 national members, and one transnational member with a total membership in excess of 15 000. An observer category is provided to give personal status to groups or organizations considering formal affiliation.

## Objectives

- To reflect the interests and initiatives of the affiliated organizations.
- To generate and disseminate information of interest to the medical and biological engineering community and international organizations.
- To provide an international forum for the exchange of ideas and concepts.
- To encourage and foster research and application of medical and biological engineering knowledge and techniques in support of life quality and cost-effective health care.
- To stimulate international cooperation and collaboration on medical and biological engineering matters.
- To encourage educational programmes which develop scientific and technical expertise in medical and biological engineering.

## Activities

The IFMBE has published the journal *Medical and Biological Engineering and Computing* for over 34 years. A new journal *Cellular Engineering* was established in 1996 in order to stimulate this emerging field in biomedical engineering. In *IFMBE News* members are kept informed of the developments in the Federation. *Clinical Engineering Update* is a publication of our division of Clinical Engineering. The Federation also has a division for Technology Assessment in Health Care.

Every three years, the IFMBE holds a World Congress on Medical Physics and Biomedical Engineering, organized in cooperation with the IOMP and the IUPESM. In addition, annual, milestone, regional conferences are organized in different regions of the world, such as the Asia Pacific, Baltic, Mediterranean, African and South American regions.

The administrative council of the IFMBE meets once or twice a year and is the steering body for the IFMBE. The council is subject to the rulings of the General Assembly which meets every three years.

For further information on the activities of the IFMBE, please contact Jos A E Spaan, Professor of Medical Physics, Academic Medical Centre, University of Amsterdam, PO Box 22660, Meibergdreef 9, 1105 AZ, Amsterdam, The Netherlands. Tel: 31 (0) 20 566 5200. Fax: 31 (0) 20 691 7233. Email: IFMBE@amc.uva.nl. WWW: http://vub.vub.ac.be/~ifmbe.

## IOMP

The IOMP was founded in 1963. The membership includes 64 national societies, two international organizations and 12 000 individuals. Membership of IOMP consists of individual members of the Adhering National Organizations. Two other forms of membership are available, namely Affiliated Regional Organization and Corporate Members. The IOMP is administered by a Council, which consists of delegates from each of the Adhering National Organization; regular meetings of Council are held every three years at the International Conference on Medical Physics (ICMP). The Officers of the Council are the President, the Vice-President and the Secretary-General. IOMP committees include: developing countries, education and training; nominating; and publications.

### Objectives

- To organize international cooperation in medical physics in all its aspects, especially in developing countries.
- To encourage and advise on the formation of national organizations of medical physics in those countries which lack such organizations.

### Activities

Official publications of the IOMP are *Physiological Measurement, Physics in Medicine and Biology* and the *Medical Science Series*, all published by Institute of Physics Publishing. The IOMP publishes a bulletin *Medical Physics World* twice a year.

Two Council meetings and one General Assembly are held every three years at the ICMP. The most recent ICMPs were held in Kyoto, Japan (1991) and Rio de Janeiro, Brazil (1994). Future conferences are scheduled for Nice, France (1997) and Chicago, USA (2000). These conferences are normally held in collaboration with the IFMBE to form the World Congress on Medical Physics and Biomedical Engineering. The IOMP also sponsors occasional international conferences, workshops and courses.

For further information contact: Hans Svensson, PhD, DSc, Professor, Radiation Physics Department, University Hospital, 90185 Umeå, Sweden. Tel: (46) 90 785 3891. Fax: (46) 90 785 1588. Email: Hans.Svensson@radfys.umu.se.

# CONTENTS

PREFACE TO THE FIRST EDITION — xiii

PREFACE TO THE SECOND EDITION — xv

ACKNOWLEDGMENTS — xvii

1 GENERAL DESCRIPTION OF A LINEAR ACCELERATOR AND ITS COMPONENT SYSTEMS — 1

2 BASIC PROCESSES IN ELECTRON ACCELERATION—THE ACCELERATING WAVEGUIDE — 10
  2.1 Introduction — 10
  2.2 Waveguide theory — 11
      2.2.1 The travelling wave accelerator — 13
      2.2.2 The standing wave accelerator — 17
  2.3 Control of the energy of the accelerated electrons — 19
      2.3.1 Travelling wave accelerators — 19
      2.3.2 Standing wave accelerators — 20
      2.3.3 Coupled and multipass linear accelerators — 21
      2.3.4 Electron energy and electron beam current — 22
      2.3.5 Magnitude of electron beam current — 23
  2.4 Comparison of travelling wave and standing wave accelerators — 23
  2.5 Construction of the accelerator waveguide — 24

3 THE MICROWAVE SYSTEM AND ITS HIGH-VOLTAGE SUPPLIES — 27
  3.1 The microwave circuit — 27
  3.2 Microwave source requirements — 29
      3.2.1 The magnetron — 30
      3.2.2 The klystron — 33
  3.3 Microwave power transmission — 34
      3.3.1 Rectangular waveguides — 34

|       |       | 3.3.2 | Transition sections and windows | 35 |
|---|---|---|---|---|

- 3.3.2 Transition sections and windows — 35
- 3.3.3 Isolators — 35
- 3.3.4 Phase shifters — 36
- 3.3.5 Four-port circulators — 37
- 3.3.6 Four-port hybrid couplers — 37
- 3.3.7 RF loads — 37
- 3.3.8 Rotary joints — 38
- 3.4 Control of microwave frequency — 38
  - 3.4.1 Magnetron driven machines — 38
  - 3.4.2 Klystron driven machines — 42
- 3.5 The modulator — 43
  - 3.5.1 Waveforms — 45

## 4 THE VACUUM, COOLING AND ANCILLARY SYSTEMS — 48
- 4.1 The vacuum system — 48
  - 4.1.1 Ion pumps — 49
  - 4.1.2 Roughing pumps and operation of the vacuum system — 50
- 4.2 High-pressure gas systems — 52
- 4.3 The water cooling system — 52

## 5 THE ELECTRON BEAM (ITS PRODUCTION AND TRANSPORT) — 55
- 5.1 Beam optics — 55
  - 5.1.1 Magnetic dipoles — 56
  - 5.1.2 Solenoids — 57
- 5.2 Electron guns — 58
  - 5.2.1 Diode guns — 58
  - 5.2.2 Triode guns — 61
- 5.3 Beam transport — 62
  - 5.3.1 Steering coils — 62
  - 5.3.2 Focusing coils — 63
  - 5.3.3 Beam bending — 65
- 5.4 The complete electron accelerating system — 69

## 6 THE TREATMENT HEAD — 71
- 6.1 The x-ray target — 75
- 6.2 The beam flattening filter — 75
- 6.3 Primary collimator and beam defining collimators — 78
  - 6.3.1 Symmetric collimators — 79
  - 6.3.2 Independent collimators — 79
  - 6.3.3 Multileaf collimators — 80
- 6.4 Wedge filters — 83
  - 6.4.1 Removable wedges — 84

|  |  |  | |
|---|---|---|---:|
| | 6.4.2 | Universal or motorized wedges | 85 |
| | 6.4.3 | Dynamic wedges | 87 |
| 6.5 | Electron beam production | | 88 |
| | 6.5.1 | Scattered and collimated electron beams | 88 |
| | 6.5.2 | Scanned electron beams | 90 |
| 6.6 | Multipurpose treatment machines | | 91 |
| 6.7 | Scales and display of the treatment head settings | | 92 |
| 6.8 | Circuits connected to the treatment head | | 92 |

## 7  THE DOSE MONITORING AND CONTROL SYSTEM — 94

- 7.1 The beam monitor — 94
- 7.2 Monitoring and control of beam uniformity—x-rays — 96
  - 7.2.1 Control of beam steering — 98
  - 7.2.2 Control of beam energy — 99
  - 7.2.3 A further note on beam uniformity — 100
  - 7.2.4 Control of machines without bending or using achromatic bending — 100
- 7.3 Monitoring and control of dose rate and beam uniformity—electrons — 101
- 7.4 Dose and dose rate monitoring — 101
  - 7.4.1 Principles of a single-channel dose monitor — 101
  - 7.4.2 A dual-channel dose monitor — 104

## 8  BEAM DIRECTION AND BEAM SHAPING DEVICES — 108

- 8.1 The main optical beam — 108
  - 8.1.1 The relation of the x-ray field to the optical field — 108
  - 8.1.2 The relation of the electron field to the optical field — 111
- 8.2 Mechanical pointers — 112
- 8.3 Optical pointers — 113
  - 8.3.1 The optical SSD scale — 114
  - 8.3.2 Axis lights — 115
  - 8.3.3 Optical back pointers — 116
- 8.4 Beam shaping devices — 116
  - 8.4.1 X-ray fields — 116
  - 8.4.2 Moulded shadow blocks — 118
  - 8.4.3 Electron fields — 119
- 8.5 Compensators — 119
- 8.6 General comment — 120

## 9  MECHANICAL SYSTEMS — 121

- 9.1 The isocentric mounting — 121
  - 9.1.1 Treatment volume centred on the isocentre — 122
  - 9.1.2 Beam entrance centred on the isocentre — 123
- 9.2 The gantry — 123

|       |                                                          |     |
|-------|----------------------------------------------------------|-----|
|       | 9.2.1 The drum mounting                                  | 125 |
|       | 9.2.2 The pendulum mounting                              | 126 |
|       | 9.2.3 Comparison of drum and pendulum mountings          | 128 |
| 9.3   | Patient support systems                                  | 128 |
|       | 9.3.1 Systems using a ram lifting mechanism              | 129 |
|       | 9.3.2 Systems using a scissor lifting mechanism          | 131 |
|       | 9.3.3 The patient couch                                  | 131 |
|       | 9.3.4 A novel patient support system                     | 133 |
| 9.4   | Movement control systems                                 | 134 |
| 9.5   | Scales for mechanical systems                            | 136 |
| 9.6   | Patient safety                                           | 136 |

## 10 CONTROL AND INTERLOCK SYSTEMS — 138
- 10.1 Machine controls and interlocks — 139
- 10.2 Safety interlocks — 139
- 10.3 Treatment controls and interlocks — 141
  - 10.3.1 Treatment prescription — 141
  - 10.3.2 Operation of treatment controls — 145
  - 10.3.3 General comment on machine operation — 147
- 10.4 Control and interlock circuits — 147
  - 10.4.1 Interlock circuits for radiation control — 147
  - 10.4.2 Analogue control circuits — 150
- 10.5 Computer control — 152
  - 10.5.1 Machines interfaced to computers — 153
  - 10.5.2 Machines controlled by computers — 154
  - 10.5.3 A novel application of computer control — 159
- 10.6 Control consoles — 161
- 10.7 General comment — 163

## 11 TREATMENT VERIFICATION — 164
- 11.1 Verification of machine operating conditions — 164
- 11.2 Verification of dose delivery — 166
  - 11.2.1 Thermoluminescent dosimetry — 167
  - 11.2.2 Silicon diode dosimetry — 168
- 11.3 Verification of irradiation geometry—portal imaging — 169
  - 11.3.1 Electronic portal imaging devices — 169

## 12 SPECIFICATION, PERFORMANCE AND CALIBRATION — 174
- 12.1 Mechanical systems — 174
  - 12.1.1 The treatment head — 175
  - 12.1.2 The gantry — 177
  - 12.1.3 The patient support system — 177
  - 12.1.4 Note on isocentric accuracy — 177
  - 12.1.5 Movement control — 178

| | | |
|---|---|---|
| | 12.2 Radiation output | 178 |
| |     12.2.1 Specification of the dose monitor | 181 |
| | 12.3 Properties of radiation beams | 182 |
| |     12.3.1 X-ray and electron depth dose curves | 182 |
| |     12.3.2 Isodose charts | 183 |
| | 12.4 Calibration of the dose monitor | 186 |
| |     12.4.1 X-ray calibration standardization | 187 |
| |     12.4.2 Electron beam calibration | 189 |
| |     12.4.3 Calibration of the dose monitor for a multipurpose linear accelerator | 189 |
| | 12.5 Specification of dose rate | 189 |
| |     12.5.1 Dose rate and arc therapy | 191 |
| | 12.6 Use of a radiation beam to demonstrate the position of the isocentre | 191 |
| | 12.7 Use of film to show the position and size of the x-ray field | 192 |
| | 12.8 General comment | 192 |
| 13 | **RADIATION PROTECTION AND ROOM DESIGN** | **194** |
| | 13.1 Unwanted radiation | 194 |
| |     13.1.1 Leakage | 194 |
| |     13.1.2 Neutrons and induced radioactivity | 197 |
| | 13.2 The treatment room and its environment | 199 |
| |     13.2.1 Treatment room layout | 199 |
| |     13.2.2 Access to the treatment room | 201 |
| |     13.2.3 The dimensions of the treatment room | 203 |
| |     13.2.4 Detailed design of protective structures | 204 |
| |     13.2.5 Shielding data and determination of barrier thickness | 204 |
| | 13.3 Neutron shielding | 208 |
| | 13.4 Miscellaneous points about treatment room services | 209 |
| |     13.4.1 Control unit | 209 |
| |     13.4.2 Control of access | 210 |
| |     13.4.3 Warning lights and signs | 210 |
| |     13.4.4 Power supplies | 211 |
| |     13.4.5 Axis lights | 211 |
| |     13.4.6 Room lighting | 211 |
| |     13.4.7 Ventilation | 211 |
| |     13.4.8 Patient viewing and communication | 211 |
| |     13.4.9 Access for the machine | 212 |
| | 13.5 General comment | 212 |
| 14 | **ACCELERATOR OPERATION** | **213** |
| | 14.1 Commissioning | 213 |
| | 14.2 Routine radiation measurements | 215 |
| |     14.2.1 X-ray measurements | 215 |

|  |  |
|---|---:|
| 14.2.2 Electron field measurements | 217 |
| 14.2.3 Detailed routine measurements | 218 |
| 14.3 Routine maintenance and servicing | 223 |
|     14.3.1 Some servicing details | 224 |
|     14.3.2 Record keeping | 226 |
|     14.3.3 Electrical wiring | 227 |
|     14.3.4 Spares | 227 |
|     14.3.5 Long-term operation of linear accelerators | 228 |
|     14.3.6 General comment | 231 |
| 14.4 Operating costs | 232 |
| **15 SIMULATORS AND TOMOGRAPHIC SCANNERS** | **235** |
| 15.1 Simulators | 235 |
|     15.1.1 The x-ray head | 237 |
|     15.1.2 The image intensifier | 238 |
|     15.1.3 The patient support system | 239 |
|     15.1.4 The gantry | 239 |
|     15.1.5 Control systems | 239 |
|     15.1.6 Simulator operation | 240 |
|     15.1.7 Design of the simulator room | 240 |
| 15.2 Computed tomography | 241 |
|     15.2.1 X-ray CT scanners | 241 |
|     15.2.2 Production of computed tomographs—simulator based CT scanning | 242 |
|     15.2.3 Magnetic resonance imaging—MR scanning | 245 |
|     15.2.4 CT simulation—virtual simulation | 245 |
| **16 CONTEMPORARY DEVELOPMENTS** | **247** |
| 16.1 Dynamic therapy | 247 |
|     16.1.1 X-ray rotation therapy | 247 |
|     16.1.2 Electron rotation therapy | 249 |
|     16.1.3 Conformal therapy | 249 |
| 16.2 Stereotactic radiosurgery | 253 |
| **17 CONCLUSION** | **256** |
| **REFERENCES** | **258** |
| **INDEX** | **265** |

# PREFACE TO THE FIRST EDITION

Electron linear accelerators have been in clinical use since the early 1950s, either to produce fast electron beams or to generate x-rays for radiation therapy, and machines of this type have become the mainstay of most radiotherapy departments. Although there is no limit, in principle, to the electron energy that can be achieved by this technology, apart from the sheer length of the accelerating structure itself, the useful electron energy range for radiotherapy has proved to be about 4–40 MeV. There are no clinical or technical considerations which define either of these limits in a precise way. In broad terms, the lower limit corresponds to the radiation quality for either fast electron or x-ray beams at which the clinical advantages of megavoltage therapy become manifest. It is also about the minimal electron energy for which the technology is suited. The upper limit is roughly related to the size of the human body and this determines the maximum penetration required.

The clinical application of linear accelerators imposes stringent requirements on the machine design which are outlined below.

(i)   The radiation beam must be well defined, and variable in size.
(ii)  Inside the beam the radiation dose pattern should be either uniform or non-uniform in a precise and controllable way.
(iii) The radiation dose pattern should be stable, not only over the period of treatment, but over the useful life of the equipment.
(iv)  Requirement (iii) implies that the energy, intensity distribution, position and direction of the electron beam all need to be controlled at the point where it strikes the x-ray target or where it emerges from the machine to become a fast electron beam.
(v)   The radiation dose delivered to the patient has to be accurately monitored.
(vi)  The radiation beam has to be steerable, that is, the beam should be movable so that it can be applied in any desired position and direction at the patient.
(vii) In order to utilize the steerable beam, the position of the patient support system (usually, but not always, a couch) needs to be movable in three dimensions with high precision.

(viii) Since radiotherapy treatments are usually delivered in a predetermined number of fractions, over a period of weeks, the equipment as a whole must operate at a very high level of reliability. Reliability is also important because the high capital cost of linear accelerators requires good utilization of the equipment. Saying this in another way, reliability allows the treatment of a large number of patients.

(ix) A high standard of electrical, mechanical and radiation safety is essential.

These requirements and their consequences are discussed in the succeeding chapters.

The basic technology for generating and handling microwave radiation was developed for use in radar just before and during the Second World War and is fundamental to the operation of electron linear accelerators. When they were first brought into clinical use about thirty years ago they proved stable and highly reliable devices, mainly because of the large research and development effort that had been made to produce reliable microwave components for radar apparatus.

Linear accelerators for clinical use are now manufactured by several companies in different countries and there are obviously very substantial differences in the way these machines are designed and operated. Furthermore, details of the electronic circuitry are liable to change from year to year even for the same make and model. Because of these differences and possible changes this book deals with electronic circuits at fairly detailed block diagram level rather than at component level, in order to bring out the principles that are involved. Detailed information about particular machines is, of course, normally supplied in the manufacturer's manual.

**D Greene**
Manchester 1985

# PREFACE TO THE SECOND EDITION

While the basic technology for accelerating electrons has not changed since the previous edition, there have been significant changes in the ability to control the size and shape of the radiation field and to match it to the needs of individual patients. There have also been significant changes resulting from the use of computer technology to monitor and control the operations of the equipment and, partly as a result of this, multipurpose machines providing a range of x-ray and electron beam qualities are more widely used.

Methods of calibrating dose monitoring systems have changed and, in principle at least, have become simpler since the national standardizing laboratories have provided absorbed dose standards rather than the previous exposure standards.

These and other changes have required quite a lot of new material in this edition. In addition there has been some rewriting of existing material, partly to interpose the incorporation of the newer technology and partly, we hope, to improve clarity.

Again, as was said in the preface to the first edition, different manufacturers have adopted different ways of using new technology and they cannot all be included. The attempt here is to offer descriptions of particular equipment only to bring out the principles involved rather than to endorse any particular solution.

**D Greene and P C Williams**
Manchester 1997

# ACKNOWLEDGMENTS

We are indebted to many colleagues from the Christie Hospital Medical Physics Department, both past and present, who have helped in the preparation of this book and its earlier edition. Particular thanks are due to Mr P M Fallas and Mr C McDonagh, two of the engineers, and Mr T J Jordan, one of the physicists, who share our interest in this field.

Mrs M R Clayton helped us to produce the manuscript and Mr P Chantry, from the Department of Medical Illustration, prepared the many new diagrams required. We are pleased to acknowledge the information provided by several commercial companies, in particular Elekta Oncology Systems (previously Philips Medical Systems—Radiotherapy) and Varian Oncology Systems, who kindly supplied photographs.

# CHAPTER 1

# GENERAL DESCRIPTION OF A LINEAR ACCELERATOR AND ITS COMPONENT SYSTEMS

In strict terminology the 'linear accelerator' is only that part of the radiation treatment machine in which electrons are accelerated up to the required energy, which may be from 4 MeV for a low-energy machine to a few tens of MeV for a higher-energy machine. In general usage, and in this book, the term linear accelerator is used as a description of the whole system used for the delivery of radiotherapy. Figure 1.1 shows three photographs of the main parts of a linear accelerator for clinical use. Figure 1.1(a) shows the unit which carries the linear accelerator waveguide and the beam defining system as well as the patient support system. Figure 1.1(b) shows the control unit which is placed outside the treatment room and figure 1.1(c) shows the high-voltage supply and pulse modulator which, on this system, is mounted on the accelerator stand.

Figure 1.2, whose function is mainly to identify the different parts of the accelerator needed for the generation of the radiation beam, lays the system out in block form. The arrows between the components indicate the main inter-connections and flow of energy. For simplicity several essential but ancillary systems are not shown: these include control systems, interlock systems, cooling systems and the vacuum system, all of which will be described in later sections.

Electrons are produced by thermionic emission in the electron gun, which injects a pulse of electrons into the electron accelerator. This is a waveguide structure in which energy is transferred to the electrons from the RF fields set up by microwaves, typically at a frequency of 3000 MHz (100 mm wavelength in free space). The microwave radiation is supplied in short pulses, a few microseconds long, and this is generated by applying high-voltage pulses of about 50 kV from the pulse modulator to the microwave generator, which is most commonly a magnetron valve. In some higher-energy accelerators a klystron valve is used as the microwave power source.

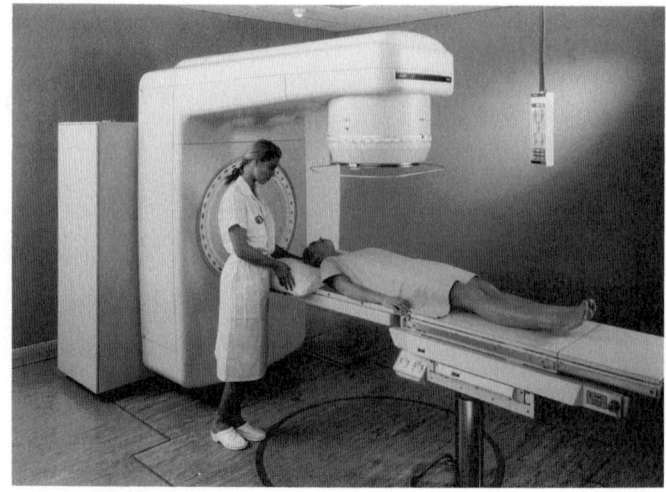

(a)

(b)

**Figure 1.1.** *(a) A general view of a low-energy linear accelerator showing the rotating structure which carries the linear accelerator waveguide and the beam defining system as well as the patient support system. (Courtesy of Elekta Oncology Systems.) (b) The control area outside the treatment room, including visual display units on the control desk and machine control unit under the desk.*

The electron gun and microwave source are pulsed so that the high-velocity electrons are injected into the accelerating waveguide at the same time as it is energized by the microwaves. The electron gun and accelerating waveguide system have to be evacuated to a low pressure such that the

(c)

**Figure 1.1.** (c) *The high-voltage supply and pulse modulator, mounted on the accelerator stand. A high-voltage rectifier stack is on the right and the thyratron valve in the centre of the photograph. (Courtesy of Elekta Oncology Systems.)*

mean free path of electrons between atomic collisions is long compared with the electron path through the system. At this stage it may be useful to discuss why the electrons are delivered in pulses. The energy that an electron can acquire from the microwave frequency field in the waveguide depends on the amplitude of the electric field, which in turn depends on the instantaneous microwave power. Even for low-energy accelerators the power required is several megawatts but, as a result of thermal and other constraints, available microwave power sources cannot be operated at these power levels continuously nor could the cooling system for the waveguide dissipate the power losses. However if they are pulsed at high power levels, with a duty cycle of around 0.1%, acceptable mean power can be delivered to pulses of electrons which can then be accelerated up to the necessary high energies. Pulses are typically of 4 $\mu$s duration and are delivered at a pulse repetition frequency (PRF) of 250 Hz.

The accelerating electrons tend to diverge, partly by mutual Coulomb

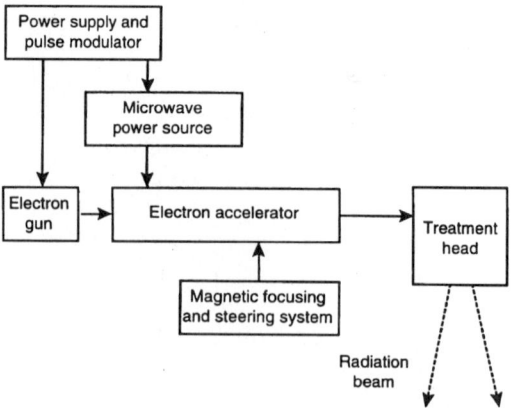

**Figure 1.2.** *A block diagram, illustrating the different parts of the accelerator needed for the generation of the radiation beam.*

repulsion but mainly because the electric fields in the waveguide structure have a radial component. However this divergence can be limited and the electrons focused back on to their straight path by the use of a coaxial magnetic focusing field. This field is generated by coils which themselves are coaxial with the accelerating waveguide. There are also additional steering coils which can be used to guide the electron beam so that it emerges from the accelerator structure at the required position and direction. The 'magnetic focusing and steering systems' are shown in figure 1.2 as a single system but are in fact distributed along the path of the electron beam from the low-energy end of the accelerator to the treatment head.

The electron accelerator delivers pulses of high-energy electrons into the treatment head, where the useful radiation beam is produced.

When the machine is being used as an x-ray generator the electron beam is directed onto an x-ray target where it is stopped with the emission of x-rays by bremsstralung. The target is therefore the primary source of x-rays, which are then modified by other components in the treatment head including filters to control beam uniformity and the beam defining system to control the dimensions normal to the beam direction. The treatment head also contains a beam monitor consisting of detectors used both to measure the dose delivered and also to provide control signals for safety interlocks and for feedback systems to the accelerator and its associated systems.

When the electron beam is to be used for treatment, it is extracted from the vacuum system through a thin window into the treatment head where it is scattered, or in some cases magnetically scanned, to give the required field coverage. This is necessary as the beam diameter is only a few millimetres as it emerges from the vacuum system. The electron beam is monitored in the same way as the x-ray beam, but in this case particular attention is paid to the detection of abnormally and dangerously high dose rates which could occur in certain fault conditions.

Some machines are designed specifically as x-ray generators but many are dual purpose with treatment heads capable of handling both x-rays and electrons. In this case elaborate mechanical and electrical arrangements are required to allow the changeover from one mode of treatment to the other. For instance, it will be necessary to have an arrangement for retracting the x-ray target and filter, inserting appropriate electron scattering foils, and to have an interlock system to verify that this has been done correctly before an electron treatment can start.

The treatment head also provides housing for an optical system simulating the treatment beams and mounting points for accessories including beam direction indication devices such as mechanical pointers which will allow precise alignment and direction of the treatment beams in relation to the patient. For electron therapy final beam collimation is carried out with an electron beam applicator which also mounts externally on the treatment head.

The temperature of certain components in the system is critical for efficient operation. In particular the temperature of the accelerating guide structure and the microwave valve has to be controlled because dimensional changes associated with thermal expansions will significantly change their frequency characteristics. The x-ray target also needs to be cooled. Cooling is achieved by circulating water through the critical components and a remote water to air (or in some cases water to waste water) heat exchanger. In either case thermostatic control is necessary to maintain the operating temperature within a few degrees centigrade.

The major mechanical elements of the linear accelerator are the gantry and patient systems, which serve to orientate the radiation source with respect to the patient lying on the patient support and positioning system. The control system and the safety interlocks mentioned above apply equally to these heavy and potentially dangerous mechanical systems as they do to the radiation generating and high-voltage circuits of the linear accelerator.

The three alternative arrangements which have been used for moving the radiation source with respect to a patient are shown schematically in figure 1.3. In the simplest of these, figure 1.3(a), the accelerating waveguide is supported so that the central axis of the treatment beam is in line with the electron path. In other words the accelerated electrons can travel in a straight line to the x-ray target, which is mounted directly at the output of the accelerator. The accelerator support system consists of a gantry base onto which is mounted a rotating gantry carrying the accelerating waveguide and the treatment head. The overall size of the structure in the horizontal direction is determined by the size of the patients to be treated. If the distance between the gantry and the central axis of the treatment beam is at least half the length of a patient, the radiation beam can then be brought to bear on any part of a recumbent patient as the patient can have either his feet or his head

6   *General description of a linear accelerator*

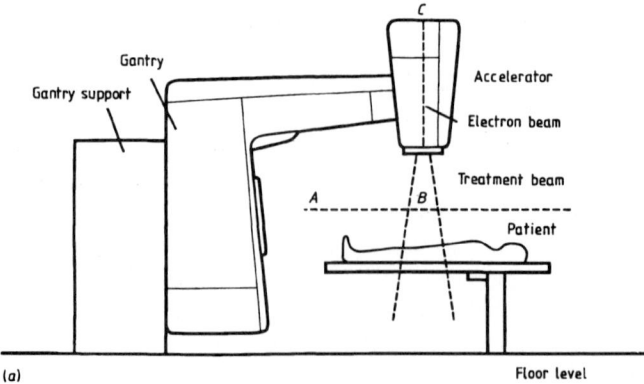

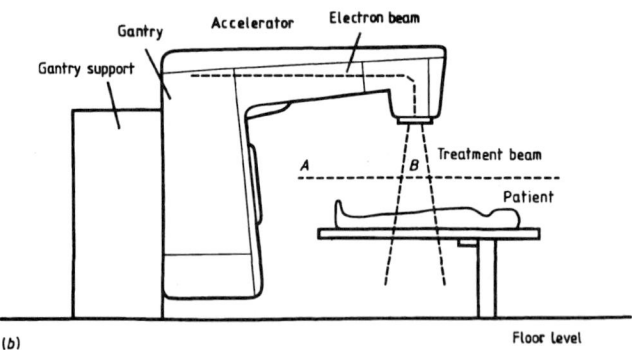

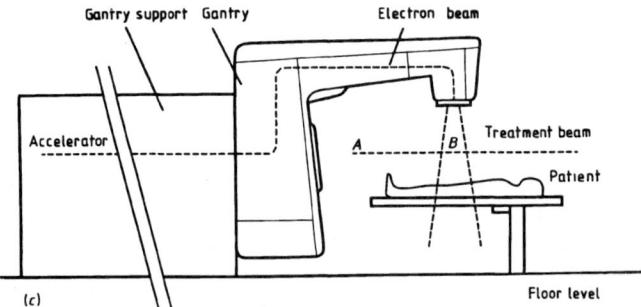

**Figure 1.3.** *The three alternative arrangements for mounting linear accelerators. (a) The accelerating waveguide is mounted so that the central axis of the treatment beam is in line with the electron path. (b) The accelerating waveguide is mounted in an arm parallel to the gantry axis of rotation AB, requiring the electron beam from the guide to be bent through approximately 90° (or 270°) onto the x-ray target or thin window. (c) The accelerating waveguide is static, and there is a rotating vacuum joint in the beam transport system consisting of flight tubes passing round the rotating section.*

towards the support structure. The overall size of the system in the radial direction, from the axis of rotation to the end of the accelerator section (BC in figure 1.3(a)) is determined mainly by the source to axis distance (SAD) which is usually 100 cm and the length of the accelerating waveguide. This is related to the electron energy and the efficiency of the accelerator, as measured by energy gained per unit length. For a travelling waveguide this is typically 5 MeV m$^{-1}$ which can be compared with 20 MeV m$^{-1}$ for a standing waveguide. In practice this means that the arrangement shown in figure 1.3(a) is practical only for linear accelerators at the lower end of the energy range and particularly those employing standing waveguides. The height of the axis of rotation AB, which corresponds to the normal position of a patient being treated, is determined by the need for staff to have access to the patient. This ideally should be about 1.2 m above floor level but is sometimes slightly higher to allow full rotation of the in line waveguide machine under the patient.

The arrangement shown in figure 1.3(b) places the accelerating waveguide in an arm approximately parallel to the gantry axis of rotation AB. This requires the electron beam from the guide to be bent through approximately 90° (or 270°) onto the x-ray target or thin window by the use of a magnetic field. The strength and distribution of this magnetic field is then critical as it determines the position of the focal spot on the x-ray target. There are several alternative designs of such bending magnets, from the simplest 90° to more complicated achromatic systems which offer advantages in beam positional control and stability. The common feature is that this arrangement makes it possible to keep the radial dimension of the rotating structure down to about 1.25 m and to rotate the treatment machine through 360°, even where relatively high electron energies are required.

A variation on the horizontal arrangement is to mount the arm and accelerator so that it passes through a drum type gantry rather than the pendulum type shown in the diagram. This allows an accelerating guide several metres long to be mounted without an excessively long arm.

There is no limit to the length of the accelerating guide that can be accommodated in the structure shown in figure 1.3(c). Here the accelerating waveguide is static, and there must be a rotating vacuum joint between it and the flight tubes passing round the rotating section. This system requires three magnetic fields to guide the accelerated electrons into the treatment head. There have been very few linear accelerators built in this configuration but it has been used in conjunction with the Microtron, a class of electron accelerator in which electrons are accelerated in steps of 0.5 MeV during each circular orbit in a magnetic field.

In each of these arrangements it must be possible to rotate the gantry under manual control so that direction from which the treatment beam is to be incident on each patient can be set up. It is also necessary to be able to rotate the gantry at predetermined speeds for rotation therapy. Both these

8      *General description of a linear accelerator*

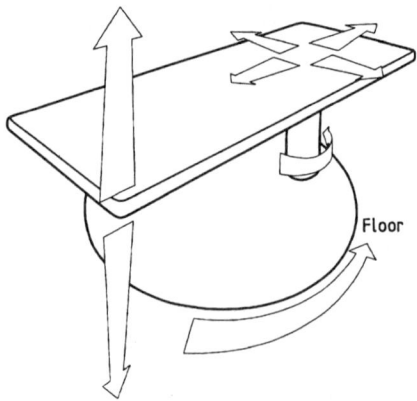

**Figure 1.4.** *A schematic diagram of the patient support system. The arrows indicate the movements of the system.*

needs can be met by using a servo controlled drive system whereby the speed of rotation may be continuously variable under manual control, whilst for rotation or dynamic therapy the speed of rotation may be accurately controlled in proportion to the dose rate so that the correct dose is delivered from each direction.

The patient support system (figure 1.4), carrying the treatment couch, must provide vertical, longitudinal and lateral movements with respect to the long axis of the couch top, and rotation about at least one vertical axis. In the system shown, the couch may be rotated about its support pillar, and the turntable carrying the whole system can also rotate about a vertical axis. Normally all the movements are motor driven but it is convenient to have the horizontal and rotational movements driven through a clutch which, when released, allows for very rapid manual positioning of the couch top. The patient is usually treated in a recumbent position on a couch supported by this mechanism, but provision may be made for replacement of the couch by a treatment chair. To provide all these facilities and movements with the required degree of precision, the patient support system is in itself quite a substantial and complex mechanical structure.

The range of technology in the system just outlined is astonishingly wide. The electron accelerating system requires relatively high-power electronics, handling peak powers of several megawatts, whilst at the other extreme the dose monitoring system has to measure currents of the order of $10^{-12}$ A.

Although the basic principles of linear accelerators have not changed since their invention in the 1950s there have been many significant developments. This is particularly true for the electronic control systems which have evolved very rapidly from valve to transistor to integrated circuits and currently into microprocessor systems. The need for and benefits from these developments, which are likely to continue, will become clear in the following chapters.

The mechanical elements in this machine design, as outlined, are basically required to place the entrance and exit points of the treatment field on a patient to an accuracy better than 2 mm. The systems that do this, the gantry, the patient support system and the beam defining system, all call for high-quality mechanical engineering.

In looking at the system as a whole, it is important to remember that in a clinical environment cleanliness, safety and appearance are essential design considerations. To at least some extent all these three functions depend on the design of machine covers which, hopefully, also help to make the equipment acceptable to patients.

# CHAPTER 2

# BASIC PROCESSES IN ELECTRON ACCELERATION—THE ACCELERATING WAVEGUIDE

## 2.1. INTRODUCTION

The idea of using an alternating voltage to accelerate charged particles in a straight path dates back to the late 1920s. A series of conducting tubes is connected to an alternating voltage supply as illustrated in figure 2.1. A positively charged particle travelling along the axis of the tubes in the diagram can then be accelerated from left to right if it passes through the gap between the first and second tubes when the polarity of the voltage is as shown. If the time required for the particle to pass through the second tube is equal to one half cycle of the alternating supply, then the electric field between the second and third tubes will be in the direction to accelerate the particle further. This process can then give energy to the particle as it passes through each gap in a long series of flight tubes. However the lengths of successive tubes must be increased so that the transit time of the accelerating particle remains equal to the half period of the alternating supply. A system using a radiofrequency supply and medium-atomic-number particles can form the basis for a practical linear accelerator, as demonstrated by Wideroe (1928) and Sloan and Lawrence (1931).

This technology is not practicable for accelerating electrons because the high velocity attained by these very light particles would require the use of

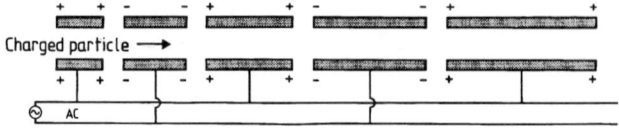

**Figure 2.1.** *Use of an alternating voltage supply to accelerate a charged particle through a series of flight tubes.*

excessively long flight tubes where a radiofrequency (RF) supply is used. The successful application of this technology to the acceleration of electrons had to wait until high-powered microwave generators, operating at higher frequencies, had been developed in the 1940s. At these frequencies, where the wavelengths become comparable to the dimensions of the flight tubes, the accelerator structure has to be developed as a waveguide system, as does the transmission line which delivers the necessary power (Fry *et al* 1947, Miller 1954, Karzmark and Pering 1973).

## 2.2. WAVEGUIDE THEORY

It would be out of place here to give a detailed exposition of standard waveguide theory, as it is readily available in many, more specialized, textbooks (see, e.g. Mooijweer 1971, Elliot 1972, Connor 1982). However, a brief outline may be helpful in showing how a waveguide structure can be used as a linear accelerator for electrons.

If an electromagnetic wave is transmitted between conducting surfaces, then it is reflected at these surfaces as shown in figure 2.2(a). These waves interfere, and energy will only be transmitted along the axis if the instantaneous interference patterns between the reflecting surfaces are coherent. This can happen only if the path length between reflections (AB in figure 2.2(a)) is an integral number of wavelengths. Under these circumstances, the instantaneous electric field pattern for (a) is as shown in (b), while (c) shows the electric field pattern along the axis of symmetry. This discussion has the following consequences.

(i) For a particular separation of the conducting planes, the length AB can meet the condition of being an integral number of wavelengths for a number of different angles between AB and the axis of symmetry; this defines the 'modes' by which the system can transmit energy.

(ii) Such modes are only possible if the wavelength is less than a cut-off wavelength, which is related to the separation of the conducting planes. This description of the propagation of waves in a guide follows that by Terman (1943).

(iii) As the waves progress between the conducting planes by a series of reflections the electric field, which would be normal to the direction of propagation in free space, now has an axial component, which we will see can be used to accelerate electrons.

Practical waveguides are somewhat more complicated than the pair of conducting planes described above. They are metallic tubes of rectangular or circular cross section in which the waves must satisfy Maxwell's equations and in particular the boundary conditions at the metallic walls. Thus there can be no tangential component of the electric field where it meets the waveguide wall and no normal component for the magnetic field at the wall.

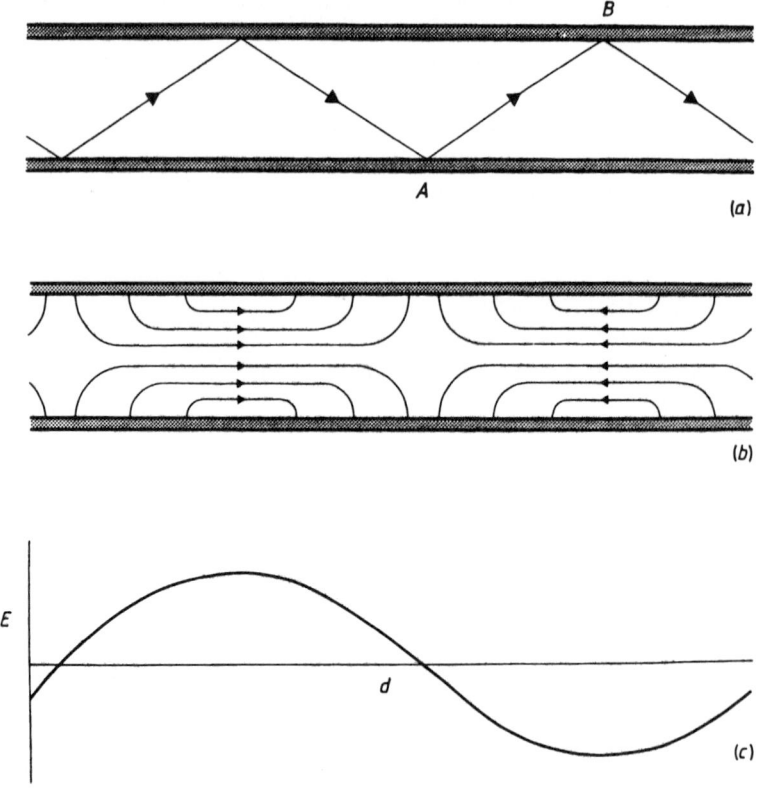

**Figure 2.2.** (a) The central ray of an electromagnetic wave being reflected between parallel conducting surfaces. (b) Instantaneous electric field distribution when distance AB in (a) is one wavelength. The arrows indicate the direction of the electric field. (c) Instantaneous electric field distribution along the axis of symmetry between the conducting planes. E is the electric field strength and d the distance along the axis.

The simplest waveguides are rectangular in section and are used in linear accelerators to transmit power between the separate parts of the microwave system, most importantly between the microwave source and the accelerating structure.

Wave transmission modes are classified by the field patterns set up inside the waveguide. There are two basic classifications: the transverse electric (TE) modes and the transverse magnetic (TM) modes. These arise as it can be shown that if the magnetic field has a component in the direction of propagation then the electric field is purely transverse (TE) and the converse is also true (TM). Further classification depends on the number of half-period variations of the fields along each axis of the guide. Although rectangular waveguides can transmit power in many modes (corresponding to the complexity of the field patterns in the waveguide) the dominant mode, which has the highest cut-off wavelength, is used as this mode is usually subject to the lowest losses.

Linear accelerators operate in the S band with a nominal frequency of 3 GHz and a nominal wavelength in free space of 10 cm. Rectangular waveguides for this band have standard dimensions ($a \times b$) of 3.4 cm × 7.2 cm and the cut-off wavelength for the dominant $TE_{0,1}$ mode is given by:

$$\lambda_0 = 2b = 14.4 \text{ cm}.$$

All other modes have cut-off wavelengths below 10 cm and are therefore not allowed.

### 2.2.1. The travelling wave accelerator

In order to accelerate electrons along the axis of a waveguide it is clearly necessary to employ a waveguide with an axial electric field, i.e. operated in one of the transverse magnetic field (TM) modes. A circular waveguide has a $TM_{0,1}$ mode with electric field patterns, in the radial plane, similar to those shown in figure 2.2(b). If we now consider an electron moving along the central axis then it can either be accelerated or decelerated according to its position in the field. In so far as the system is transmitting energy from left to right the field pattern will be moving in the same direction. If we arrange that the electron is always moving at the same velocity as the field, then an electron initially placed in the field in a position such that it receives a force in the left to right direction will receive energy continuously from the field. This process implies that it is necessary to control the rate at which microwave energy is transmitted along the waveguide, and that the field velocity can be increased to keep it in step with the velocity of the accelerating electron.

Although microwaves travel at $c$, the speed of light, in free space the velocity of propagation in a waveguide depends on the wavelength and the dimensions and shape of the guide. At a simple level this can be seen in figure 2.2(a): as the wave has to travel along the guide by a series of reflections, the resultant velocity along the axis must be less than $c$. This is the group velocity, $v_g$, in the waveguide which corresponds to the velocity at which energy is transmitted along the axis. It is not quite so obvious that the wavefronts, normal to the direction of propagation, move along the axis at a velocity greater than $c$. This is the phase velocity $v_p$, and although it does not represent any physical movement it is the velocity at which the electric field patterns in figure 2.2(b) and (c) move along the guide. It is therefore this velocity which must be controlled in order to keep an electron, whose velocity is a function of its energy, at the same point on the wave where it will continue to be accelerated.

14     *Basic processes in electron acceleration*

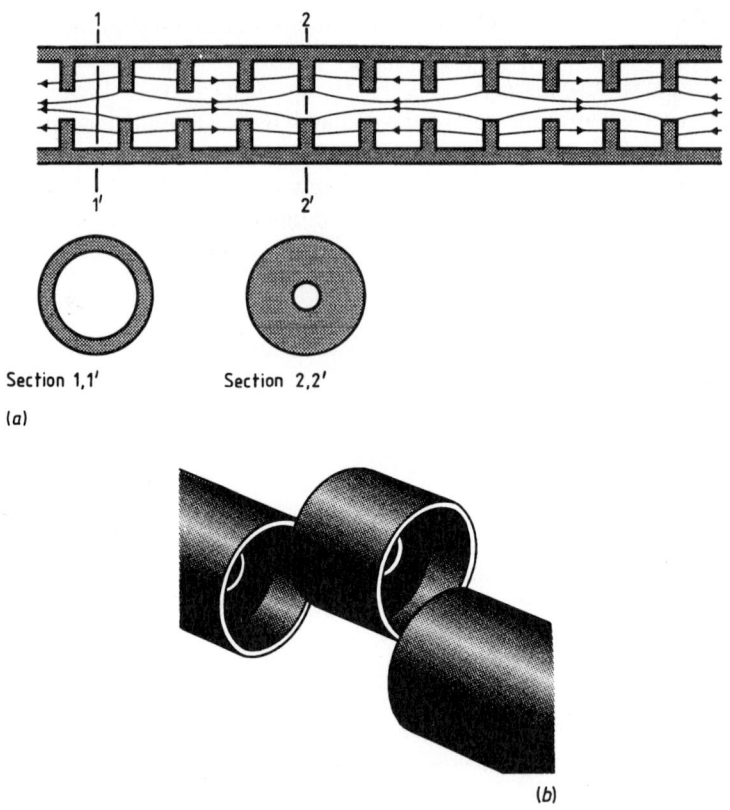

**Figure 2.3.** *(a) Sections through a disc loaded waveguide, showing arrowed electric field lines. (b) A perspective diagram of (a).*

For a circular waveguide of constant radius, $a$, the phase velocity in the $TM_{0,1}$ mode is given by

$$v_p = \frac{c}{\sqrt{1 - (\lambda/2.61\,a)^2}}. \tag{2.1}$$

From this equation it is clear that the phase velocity is always greater than $c$, but for wavelengths very much smaller than the waveguide dimension, $a$, it approaches $c$. The equation also tells us that the cut-off wavelength for a circular waveguide is $2.61\,a$ at which point the phase velocity would be infinite and the group velocity zero.

The phase velocity can be reduced by fitting the guide with conducting irises or discs, as shown in figure 2.3, which divide the waveguide into a series of coupled resonant cavities through which the microwaves propagate at a speed determined by the cavity dimensions. The instantaneous electric field distribution is also shown in figure 2.3(a) and figure 2.3(b) shows a cut-away view of a disc loaded waveguide.

The model of coupled cavities is very useful but will not be pursued here. However a simpler model can be used to explain how the irises affect the phase velocity.

The circular waveguide can be considered to be a transmission line. As with any transmission line system, the rate at which energy is transmitted is determined by the inductance per unit length, $L_0$, and capacitance per unit length, $C_0$. The phase velocity is given by

$$v_p = \frac{1}{\sqrt{L_0 C_0}}. \qquad (2.2)$$

Conducting irises in a waveguide operating in the $TM_{0,1}$ mode act as discrete capacitors in parallel with the distributed capacitance $C_0$. Thus if an iris of capacitance $C$ is added at intervals $d$ the capacitance per unit length becomes $C_0 + (C/d)$ and the phase velocity is reduced

$$v_p = \frac{1}{\sqrt{L_0 (C_0 + C/d)}}. \qquad (2.3)$$

Here both the inner diameter of the irises (determining $C$) and their spacing can be used to control the wave velocity but other considerations apply to the choice of these parameters, which also affect properties including the field strength and the attenuation of the structure which are critical to accelerator design.

A waveguide configured as an electron accelerator is shown schematically in figure 2.4(a). Electrons are injected into the waveguide from the electron gun at the same time as power is applied to the microwave generator. The main accelerating structure is at DC earth potential. A negative-going 50 kV pulse is applied to the cathode, so that electrons at 50 keV are injected into the guide.

The accelerating guide is designed such that a wave travelling from the left-hand end will start at about $0.4\,c$ corresponding to the injection energy. The velocity increases to $c$ in about the first 30 cm of the guide length in which the electrons become relativistic, and then continue travelling at this velocity until they leave the accelerating guide.

The instantaneous waveform along the long axis of the guide is as shown in figure 2.4(b), and electrons will be distributed along the guide axis such that their positions in relation to the electric field component of the wave are as shown by the points on the wave. Consider an electron at point A. This will receive a force from the electric field to make it accelerate in the direction shown. An electron behind this one on the wave pattern at B will receive a greater acceleration, and will therefore move forward in phase to

16     Basic processes in electron acceleration

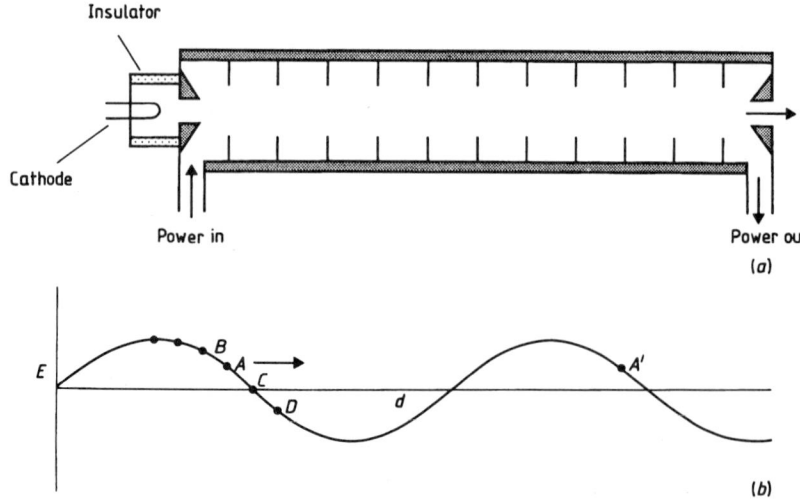

**Figure 2.4.** (a) A travelling wave accelerator. (b) A travelling wave representing the axial electric field; E is the electric field strength and d the distance along the axis. Points A–D indicate positions of electrons on the wave.

overtake that at A. An electron ahead of point A on the wave will receive a smaller accelerating force or even a decelerating force (point D) and will be overtaken by the electron at point A. There is thus a tendency for electrons to bunch on the wave at points such as A and A'. The exact position of the electron bunches on the wavefront depends on the field strength, i.e. on the amplitude of the wave and on the relationship between the wave phase velocity and the initial electron velocity. The first section of the waveguide, where the wave and the electrons are accelerating, is described as the 'buncher section'. At the end of this section, the electrons will have attained a velocity approaching the velocity of light and as they proceed through the guide they will still receive energy depending both on the field strength at the point on the wave at which they are bunched, and on the length of the accelerating guide. This increase in energy will be mainly expressed as a relativistic increase in electron mass.

As can be seen from figure 2.3(a), the electric field pattern in the accelerating waveguide has an axial component, which serves to accelerate the electron bunches as described, and a radial component, which tends to disperse them towards the walls. As previously mentioned this dispersion is controlled by the use of an axial magnetic field which provides a force on the diverging electrons acting in opposition to that from the radial electric field.

As the electromagnetic wave passes along the accelerating guide, it is attenuated partly by resistive losses in the guide structure itself, but mainly because energy is being transferred from the field to the accelerated electrons. It would appear at first sight that an ideal system would be one where all the microwave energy is passed to the charged particles, i.e. one where the electromagnetic wave amplitude is reduced to zero at the high-energy end of the system. In practice this is not very useful because the energy gained per unit length of guide would become steadily smaller towards the high-energy end. Since it is desirable for the design of the whole system to achieve maximum energy per unit length of guide it is necessary to allow electrons to pass out of the linear accelerator while the electromagnetic wave still has a relatively large amplitude.

In the travelling wave system the microwave radiation must enter the accelerating waveguide at the electron gun end. It must either pass out at the high-energy end or the remaining energy must be absorbed at this end, without reflection.

Microwave energy is fed into the gun end from a conventional rectangular waveguide as shown in figure 2.4(a). It is coupled into the circular section guide using a 'door knob' transformer, shown in section in figure 2.4(a). This is a mode transformer; its shape is arrived at empirically and it is circularly symmetrical about the diagram's long axis. A 'door knob' transformer is also used to extract the remaining microwave energy from the waveguide at its high-energy end into a rectangular guide. Here the energy can be absorbed in a resistive load. Two other alternatives have been used to handle this remaining energy. One is to feed it back to the input end of the guide with a suitable phase adjustment (phase shift) to reinforce the incoming power, while the other is to make the irises at the high-energy end of the guide out of a resistive material, so that the remaining microwave energy is absorbed directly.

## 2.2.2. The standing wave accelerator

If each end of an iris loaded waveguide is terminated with a conducting disc, microwave power will be reflected with a $\pi/2$ phase change at either end and standing waves can build up in the system.

The fundamental mode for such a standing wave is illustrated in figure 2.5(b), where cavities 2, 4 and 6 of (a) are nodes in the system, and 1, 3 and 5 are antinodes. The arrows then represent the direction of the force on a charged particle on the axis of cavities 1, 3 and 5. It can be seen that a particle passing through cavities 2, 4 and 6 will receive zero energy. One half cycle later the situation shown by the full curve in figure 2.5(b) will have reversed itself. If the time required for a particle to travel from cavity 1 to cavity 3 is equal to one half cycle, it can then be accelerated in the same direction in both cavities. A particle passing through the system can then receive energy in alternate cavities; in other words it can be accelerated by the standing waves in the system.

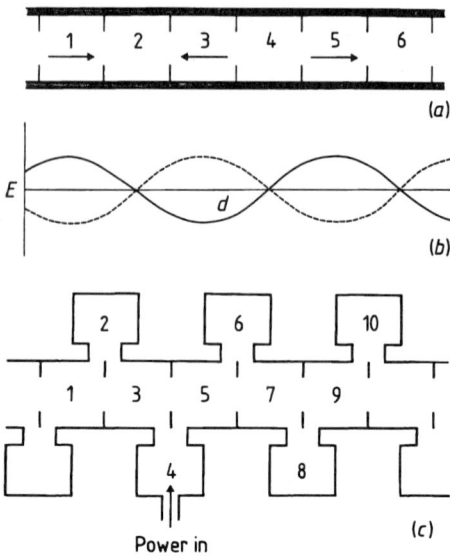

**Figure 2.5.** (a) A standing wave accelerator. (b) Standing waves, the dotted line representing the field one half cycle after the field shown by the solid line. (c) Side coupled cavities.

Since the accelerating particles receive no energy in passing through cavities 2, 4 and 6 (figure 2.5(a)), these cavities can be moved out to the side as shown in figure 2.5(c): this serves to shorten the accelerating structure and to increase the effective accelerating gradient. Side coupling is achieved by inductive coupling through peripheral slots between adjacent cavities, the advantage of inductive coupling being that the axial holes, through which the electrons must pass, can be kept small and this in turn allows more freedom in the design of the shape of the cavities, which can then be optimized to give the highest electric field. The structure as a whole needs to be fitted with end plates which provide the reflections necessary to establish standing waves. These end plates need to have axial holes in them to allow the electron beam to enter and to leave the system. When the microwave pulse is fed into the system the amplitude of the standing waves will build up until power losses in the system, including the energy passed to the electron beam, are equal to the incoming power. The amplitudes of the standing waves are constant throughout the length of the accelerator. At the end of each high-voltage pulse energizing the microwave generator and the electron gun, the oscillations in the standing wave system decay due to losses in the coupled cavities. This process needs to have a short time constant compared with the time between successive pulses, as the incoming microwave pulse may not be in phase with the previous one.

In the standing wave system the bunching function can be achieved by adjusting the axial length of the accelerating cavities, at the low-energy end, to keep the transit time through each cavity at one half cycle, as the electron velocity increases and approaches $c$. However as the electric field in a standing wave system is considerably higher than the initial field in a travelling wave system the buncher section is short, consisting of just two or three cavities for high-energy accelerators and just the first cavity for low-energy accelerators. It should be noted that the resonant frequency of the cavities depends on their diameter and not on their length. The length can therefore be chosen to match the transit time between cavities with the energy of the electrons and the freqency, so that electron bunches arrive at the right points on the standing waves at the right time.

In the travelling wave system, it is essential that the microwave power be fed into the accelerator at the gun end. This condition no longer applies in the standing wave mode, when the amplitude is constant through the whole length of the system, and so the incoming power can be supplied at any point which is convenient. It is usually fed directly into one of the coupling cavities as shown in figure 2.5(c).

## 2.3. CONTROL OF THE ENERGY OF THE ACCELERATED ELECTRONS

The output energy from a linear accelerator depends on the input power level, $P$, during each pulse, the microwave frequency and as will be discussed later, the electron beam current. In the following sections the energy is that which would be achieved at zero current, that is without beam loading. Other properties of the waveguide, such as the length and the impedance, also affect the energy but as they are fixed they are not considered in this section.

### 2.3.1. Travelling wave accelerators

In most travelling wave accelerators the peak power level is achieved instantaneously at the start of the microwave pulse and therefore the energy is constant throughout the pulse and is proportional to $\sqrt{P}$. However in some high-energy machines the residual microwave power at the end of the accelerator is fed back, in phase, to reinforce the input power, thus increasing the power in the waveguide in steps corresponding to the transit time of the microwaves through the accelerator and feedback elements. This results in small increases in energy during each pulse but reduces the effects of beam loading and the sensitivity to small changes in microwave frequency.

The wave velocity in a travelling wave accelerator is critically dependent on the microwave frequency, and two related practical consequences follow:

(i) for stable operation the frequency has to be very accurately controlled;

(ii) relatively small changes in the frequency can be used to produce wide variations in electron energy. (Methods of controlling the frequency are discussed in chapter 3.)

To explain the second of these two statements, consider again the phase bunching discussion in relation to figure 2.4(b). A change in relationship between the initial electron velocity and the wave velocity in the buncher section in the accelerating guide will alter the position of the bunches on the wavefront. Since it is this position (or phase relationship) which determines the electric field strength accelerating the electron, the final energy of the accelerated electrons would be changed. For a given accelerating waveguide this relationship makes it possible to vary the electron energy between the maximum available, down to a few MeV.

When the electron energy is varied in this way, by changing the microwave frequency, the efficiency of the system falls rapidly from that at its optimum frequency. This is tolerable when the system is being used for electron beam treatments as the accelerated electron beam current only needs to be a very small fraction of that required when the system is used as an x-ray generator. Put another way, different microwave frequencies may be used to select various electron energies for electron beam treatments, but the system can only be used as a x-ray generator at the electron energy corresponding to optimum frequency conditions.

Some linear accelerators offer two alternative radiation qualities for x-ray treatment. This is achieved by operating the system at optimum frequency, but at two different peak power levels from the microwave power generator.

### 2.3.2. Standing wave accelerators

As with the travelling wave accelerator the energy is proportional to $\sqrt{P}$. However in this case the peak power in the accelerating cavities is not established instantaneously as it takes a finite time for the standing waves to reach maximum amplitude. This filling time is typically 1 $\mu$s during which time electrons would not be accelerated efficiently. This problem can be avoided by delaying injection of the electrons, by the use of a triode gun. The consequence of this is that electron acceleration takes place during only the final 3 $\mu$s of a typical 4 $\mu$s pulse, resulting in a loss of efficiency in the use of the RF power.

The standing wave accelerator is a highly resonant system and will only work at a fixed frequency. The energy of the accelerated electrons depends mainly on the amplitude of the oscillations in the standing waveguide, and on the phase of the electron bunches in relation to the standing waves. The amplitude depends on the microwave power, while the phase can be varied by changing the energy of the electrons coming into the system from the electron gun. These two variables can be used to vary the electron energy. However in a standing wave accelerator designed to operate at a particular energy, large changes in power level result in significant reductions in beam intensity and broadening of the electron spectrum. Although a reduction in beam intensity is not a problem *per se* (as long as it is high enough) the broadening of the spectrum has serious consequences in the steering and deflection of the electron beam as will be discussed in chapter 5.

Optimal performance at more than one energy can be achieved by introducing a microwave switch to reduce the accelerating field in the high-energy part of the guide. In this region the amplitude of the standing waves will be reduced but the electrons, which are travelling at relativistic velocity, will continue to arrive in each cavity at the correct phase. The non-contact microwave switch (Tanebe and Hamm 1985) changes the resonant properties of one, or a pair, of the side coupling cavities and can be designed to reduce the amplitude in the high-energy part of the accelerator while maintaining a constant field in the buncher section and maintaining the correct phase relationships along the entire length.

### 2.3.3. *Coupled and multipass linear accelerators*

Where the requirement for a wide range of electron and x-ray energies has not been satisfied by simple linear accelerators operating as described above, several novel systems have been utilized, some of which are described here.

A coupled accelerator system is one where there are two accelerating waveguides in series. The first one, incorporating the buncher section, is operated at constant power, and delivers electrons of fixed energy into the second one. If the phase of the wave in the second accelerator is changed with respect to that in the first, the position of the electron bunches on the wave will vary and consequently their final energy will be changed. The second accelerating guide can even be phased to decelerate the electrons from the first, thus extending the range of energies available. Both accelerators can be energized from a single microwave source by the use of a power divider and a phase shifter.

In a multipass linear accelerator, such as that employed in the racetrack microtron, magnetic fields at each end of the accelerating waveguide, plus a suitable arrangement of flight tubes, can be used to recirculate the accelerated electrons back to the low-energy end of the system. Thus they can be further accelerated by passing through the guide many times. The final energy of the electrons can then be controlled and varied by selecting the number of times they pass through the system.

Either travelling wave or standing wave accelerators can be utilized in coupled and multipass systems.

One additional variant that can be mentioned was developed by the Atomic Energy Corporation Limited (AECL) in Canada. It consisted of a standing waveguide, which being symmetrical can be used to accelerate electrons in either direction if the phase of the electron bunches is correctly adjusted. In

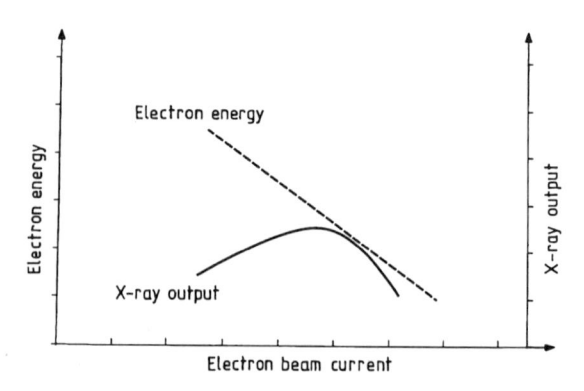

**Figure 2.6.** *The relationship between electron beam current, electron energy and x-ray output in a linear accelerator.*

this device electrons are accelerated during a first pass then extracted and turned through 180° to be accelerated up to double the energy by the second pass in the opposite direction.

These relatively elaborate systems are useful for generating variable-energy x-rays extending towards the upper end of the range considered in this book. However it should be noted that these systems have not been generally accepted by the radiotherapy community, being used mainly as research tools. The design of accelerating systems is discussed in further detail by Karzmark *et al* (1993).

### 2.3.4. Electron energy and electron beam current

For a travelling wave accelerator operating at a fixed microwave power level and frequency, the electric field amplitude is attenuated as the wave passes along the accelerating guide. This rate of attenuation depends on the shunt impedance per unit length, which along with the input power determines the amount of energy being transferred to the accelerating electrons. In other words, the rate of attenuation will be dependent on the electron beam current. Consequently the average electric field acting on an electron, and hence the energy received in passing through the system, will be reduced as the electron beam current is increased (see figure 2.6).

For a standing wave accelerator, where the electric field amplitude is constant throughout the system, the amplitude depends on the balance between incoming microwave power and losses in the microwave system including those caused by energy transfer to the electron beam. Again the average electric field acting on an electron is reduced as the electron beam current is increased, thus the electron energy is reduced.

The dependence of energy on beam current is most important at the high currents used when electron beams are used to produce x-rays; the relationship between beam current, electron energy and x-ray output is of interest. There are two conflicting effects.

(i) At constant electron energy the x-ray output is proportional to electron beam current.
(ii) X-ray production efficiency is a function of electron energy, decreasing rapidly with falling energy. This is due to an absolute reduction in efficiency compounded by a reduction in the proportion of x-rays produced in the forward direction.

Initially the first effect is dominant and x-ray output increases with electron beam current, but as the energy drops the output reaches a well defined maximum as shown in the curve in figure 2.6. This diagram is only intended to give a rough indication of the relationships concerned and any detailed version would depend on the particular accelerator concerned.

### 2.3.5. Magnitude of electron beam current

When the accelerator is used as an x-ray source the dose rate required for radiotherapy is typically 200–500 cGy† per minute at 1 m from the x-ray target. As the radiation is produced in short pulses, between 2 and 4 $\mu$s long, the mean output is determined by the electron beam current during a pulse, and by the pulse repetition frequency. At repetition frequencies of a few hundred pulses per second, the electron beam current required during each pulse turns out to be of the order of hundreds of milliamps. When the electron beam itself is used for patient treatments, then a dose rate at the patient of hundreds of cGy per minute requires the electron beam current to be about 1 mA during the pulse.

The very large difference between the electron beam currents required for electron and x-ray therapy can be a major hazard to patients. Elaborate interlock systems are thus required to ensure that the system does not run in the high-current state for electron therapy or in the low-current state for x-ray therapy.

## 2.4. COMPARISON OF TRAVELLING WAVE AND STANDING WAVE ACCELERATORS

Since both types of accelerator are still in production, albeit from different manufacturers, after many years of commercial experience, it is clear that neither type has any decisive advantage over the other. Since the standing wave system accelerates the electrons in a field of constant amplitude while

† Radiation dose is defined as energy absorbed per unit mass. The SI unit is the gray (Gy), 1 J kg$^{-1}$. The centigray (cGy) is equal to one of the old units of radiation dose (the rad; 1 rad = 100 erg g$^{-1}$) and is often used as the practical unit.

the field in a travelling wave system is attenuated as it moves along the guide, the former will give a higher electron energy in the same length of guide for a given microwave peak power level. This means that where the length of the accelerating guide is a critical design factor, a standing wave system has a definite advantage. In practice this advantage applies for accelerators operating at a few MeV and using the type of mounting outlined in figure 1.3(a). For energies above 4 or 5 MeV the guide length becomes too long for this type of mounting and the system in figure 1.3(b) has to be used. Hence the difference in accelerating guide length between the standing wave and travelling wave systems is no longer a decisive factor. If the radial dimensions of the accelerator are critical then the travelling waveguide has the advantage as the side cavities increase the radial dimensions.

Both types of accelerator require similar peak microwave power to achieve the same electron energy but the mean power is significantly higher for a standing wave accelerator because of the filling time mentioned above. Given two comparable 6 MeV accelerators the travelling wave machine could be powered by a 2 MW magnetron but the standing wave machine would require a 2.5 MW magnetron.

From the standpoint of a linear accelerator design engineer these, and other more subtle, differences are very important, but from a user's point of view the properties of the electron beams produced are so similar that they are unlikely to merit consideration in the choice of equipment. Both travelling wave and standing wave technology has proved to be effective.

## 2.5. CONSTRUCTION OF THE ACCELERATOR WAVEGUIDE

The three main requirements to be satisfied in the accelerating guide are good conductivity in the material from which it is constructed, high dimensional accuracy and high dimensional stability. The first of these is met by using pure copper as the constructional material, while the second and third require the structure to be fairly substantial with a high degree of mechanical rigidity including control of thermal expansion.

Three methods of construction can be used. The disc loaded structure of the type shown in figure 2.3(a) can be made from a pile of accurately machined sections (figure 2.3(b)), which are then clamped together. In use, a waveguide needs to be under high vacuum. This type of clamped structure cannot be vacuum tight and so the whole system has to be placed in a vacuum tank. The need for a vacuum tank is a disadvantage, both in terms of increased size and weight and because it increases the volume to be evacuated. A further disadvantage arises from the need to ensure high electrical conductivity between sections as very high currents are induced in the waveguide walls. Both these disadvantages can be avoided by brazing the sections together, so that the waveguide itself becomes the vacuum container.

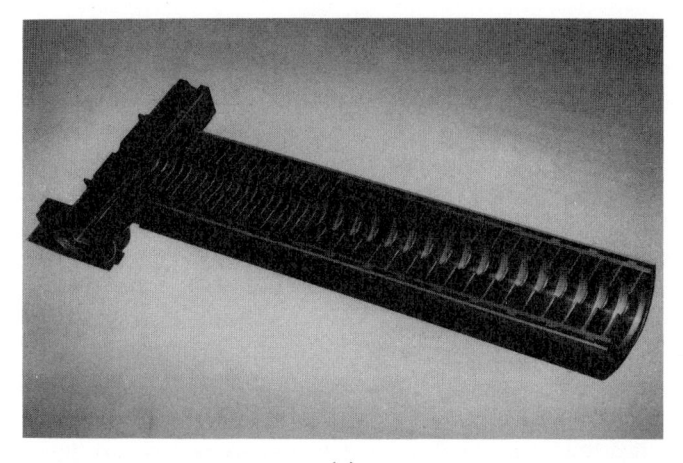

*Construction of the accelerator waveguide* 25

(a)

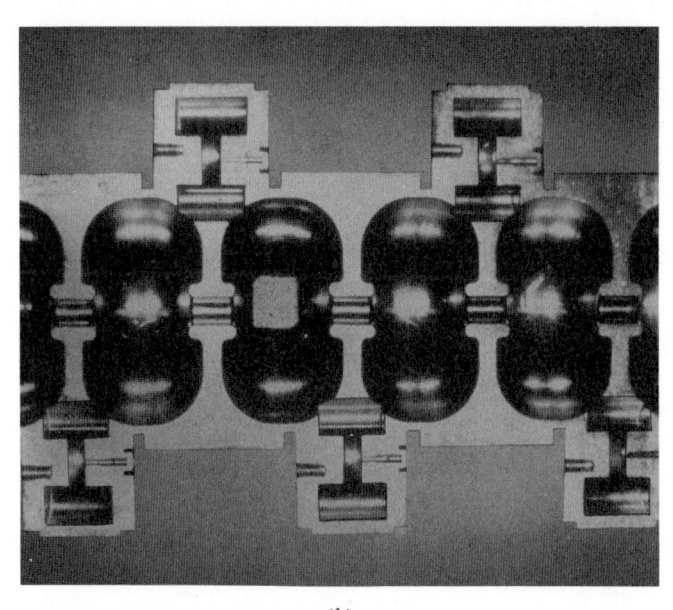

(b)

**Figure 2.7.** *(a) A photograph of a sectioned travelling waveguide complete with water jacket. (Courtesy of Elekta Oncology Systems.) (b) A photograph of a sectioned standing side coupled waveguide. (Courtesy of Varian Oncology Systems.)*

Another common method of manufacturing structures such as the one shown in figure 2.3 is electroforming. The discs of the waveguide are machined in copper. Mild steel rings of the same diameter as the discs are used as spacers to place the discs at the required distances from each other. The discs and rings are clamped together, with their outside surfaces in line, and the whole structure is placed in an electrolytic bath. Here copper is deposited on the outer cylindrical surface of the disc–ring assembly and this process is continued until the required thickness of copper has been deposited. The steel spacers are then dissolved out, leaving the waveguide structure. A photograph of a sectional waveguide structure formed in this way is shown in figure 2.7(a).

Side coupled standing waveguides are more complex in shape. The individual cavities are often electroformed, and then the whole structure is joined together by brazing. A photograph of a sectioned side coupled guide is shown in figure 2.7(b).

Mechanical tolerances in the construction of accelerating guides are of the order of 0.01 mm. They are somewhat smaller for standing waveguides than for travelling waveguides. These tolerances have to be very small in relation to the wavelength of the microwave radiation used to energize the system. On assembly, waveguides are tested at low microwave power at their design frequency and can be fine tuned by producing minor mechanical deformations in order to adjust the resonant frequency of each cavity.

Both because of resistive losses in the guide walls and because some electrons may strike the structure, the accelerator guide will heat up in operation. Consequently, thermal expansion may significantly change the dimensions. To counteract this effect, the guide is fitted with a water jacket through which temperature controlled water is circulated at a predetermined rate.

Typical outer dimensions of the travelling waveguide structure are a diameter of about 15 cm (including the water jacket) and a length between 1 and 3 m depending on the electron energy required. Such a system will weigh several hundred kilograms and has to be carefully supported to avoid deflections as it is rotated on the gantry of the treatment machine.

It should be noted that all the diagrams of accelerating waveguides in this chapter are schematic with the cavities shown pill box shaped. Although many early accelerators were this simple more recent designs have optimized the shapes of the cavities and irises in order to improve the efficiency and to concentrate the high fields on the beam axis. Cavities are often oblate spheroids and the irises have rounded edges to their inner aperture or become wider at this point so that, on the axis, the irises are short tubes rather than thin discs.

# CHAPTER 3

# THE MICROWAVE SYSTEM AND ITS HIGH-VOLTAGE SUPPLIES

Operation of the accelerating waveguide requires that it be supplied with pulses of microwave power at the appropriate peak power level and pulse repetition frequency. This chapter outlines the circuits and components which perform these functions.

## 3.1. THE MICROWAVE CIRCUIT

Three alternative microwave circuits are shown in figures 3.1, 3.2 and 3.3: the first is for a low-energy magnetron powered machine, the second for a higher-energy machine powered by a magnetron and the third for a machine powered by a klystron. Microwaves are generated by the microwave source and transferred to the accelerating structure via rectangular waveguide sections. In all cases it is necessary to prevent reflected power being returned to the source.

In the low-energy magnetron driven machine this is achieved by use of an RF isolator but in the more complicated high-energy magnetron driven machine, as shown in figure 3.2, the power passes through a four-port coupler which directs the microwaves into the accelerator, together with the in phase component of any residual power from the accelerator which is fed back via a high-power phase shifter. The component of the residual power which is not in phase with the source is diverted into a water load, a lossy section of waveguide, where it is absorbed and dissipated as heat.

The third configuration, shown in figure 3.3, is for a klystron driven machine. Power from the klystron enters a four-port circulator through port 1, and passes out through port 2 into the 'shunt tee' waveguide section. This will reflect power back into the circulator, the proportion depending on the position of the movable shunt. The reflected power then re-enters the circulator and leaves through port 3 into the accelerator. Power not reflected in the shunt tee is absorbed in water load connected to port 2.

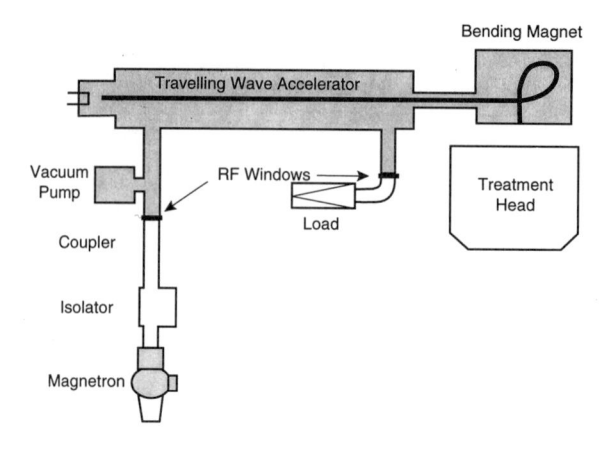

**Figure 3.1.** *A microwave circuit for a low-energy magnetron powered machine.*

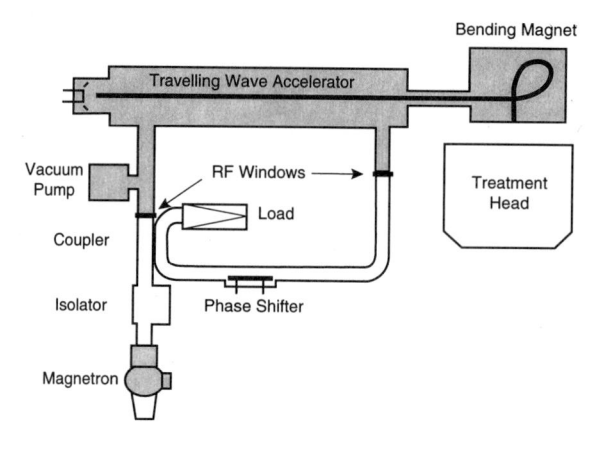

**Figure 3.2.** *A microwave circuit for a high-energy magnetron powered machine with microwave feedback.*

The power reflected from the accelerator back to port 3 is absorbed by the water load connected to port 4. The shunt on the shunt tee provides fine control of the power fed to the accelerator, its position being controlled according to the electron energy selected. As well as providing microwave power control, the four-port circulator also prevents the build-up of standing waves in the microwave transmission line by directing reflected power into the water load 4. (Note that an alternative to the use of the shunt tee for power control is to vary the power output from the klystron by running it in a highly controlled linear mode of operation.)

RF probes are inserted into the microwave circuit to provide the necessary

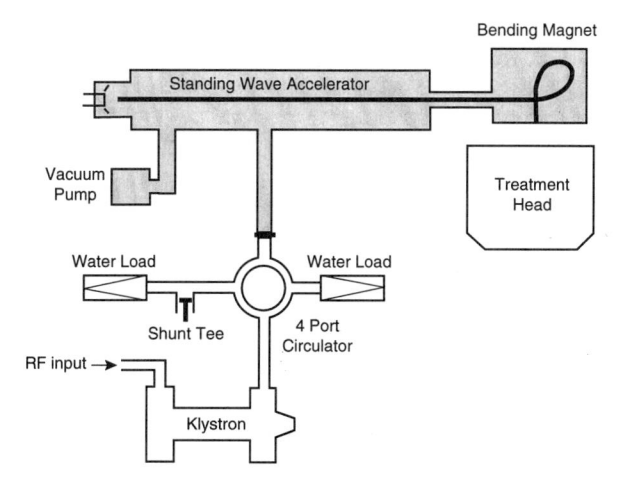

**Figure 3.3.** *A microwave circuit for a high-energy klystron powered machine.*

input to the automatic frequency control (AFC) and/or phase control systems.

## 3.2. MICROWAVE SOURCE REQUIREMENTS

Electron linear accelerators for radiotherapy are usually operated at a frequency of 3000 MHz (S band). Machines operating at three times this frequency (X band) have also been constructed. In theory the use of the higher frequency has the advantage of bringing down the dimensions of the accelerating waveguide and thus permitting the design of a more compact system. It also has the disadvantage of bringing down the dimensional tolerances by the same factor. This, plus difficulties in finding reliable microwave generators to work at the required power level, has made it impracticable to operate X band accelerators up to the present time. For very high-power research accelerators, lower-frequency (1000 MHz, L band) machines have proved to be very successful. By the same argument as above these are less compact, and for this reason, and also because S band machines have a very good performance record, they have not found any significant use as radiotherapy generators.

The microwave generators for radiotherapy accelerators are either magnetrons or klystrons. For electron energies up to about 10 MeV, a magnetron operating at a peak power of 2.5–3 MW is used. For higher-energy accelerators higher peak power levels are required. Magnetrons operating at peak power of up to 5 MW or klystrons operating at peak power of up to 7 MW have been designed to meet this requirement.

No difference in principle is involved in the choice of magnetron or klystron as power source for the accelerating waveguide; the choice is essentially one of practical and commercial convenience.

Magnetrons are physically smaller and operate at lower voltages so that they can be mounted on the rotating gantry of a linear accelerator. Klystrons are somewhat larger, operate at higher voltages and are mounted within a tank of insulating oil, which cannot be mounted on the gantry. This necessitates a rotating RF joint to transport the power to the accelerator.

*3.2.1. The magnetron*

The anode of this valve consists of a symmetrical array of cavities arranged round a cylindrical hole in a block of copper, as shown in figure 3.4(a). The cathode is an oxide coated cylinder, held concentrically in the anode, and indirectly heated by a tungsten spiral. The whole structure is placed in a uniform magnetic field with its lines of force normal to the plane of figure 3.4(a) provided by the poles of a permanent magnet (figure 3.4(b)). The magnetic field strength and anode–cathode voltage are chosen so that electrons emitted from the heated cathode follow a curved path such that they would just fail to strike the anode if the whole anode structure were at the same potential. Figure 3.4(c) shows a photograph of a magnetron, which has been cut open to show the anode–cathode structure. (A brief review of the effects of magnetic fields on electron beams is given in chapter 5.)

The anode structure forms a set of tightly coupled resonant cavities and these will oscillate in the fundamental mode such that potential differences as shown in figure 3.4(a) will alternate with the opposite configuration of + and − signs, at a frequency determined by the dimensions of the structure. An electron crossing the mouth of one of the cavities will start the system oscillating. When the valve is energized by applying a negative-going voltage pulse to the cathode relative to the anode, the electrons from the cathode circulate concentrically round the anode cathode space, and as the fields between the anode cavities starts to oscillate electrons may be either accelerated or decelerated in passing the mouth of a cavity. This will cause the electrons to bunch by the same process as described for the accelerator waveguide's buncher section. The electrons circulating in the anode space form a space charge cloud shaped like a set of rotating wheel spokes, where the 'spokes' cross the mouths of successive cavities at times when they are decelerated. The rotating electrons then give energy to the oscillating cavities, and the amplitude of the oscillations increases until the energy transferred from the electrons equals the losses in the oscillating system.

The peak power into the system is determined by the electron emission from the cathode and the applied voltage, and at 2.5 MW both anode and cathode need to be substantial structures. The anode is water cooled, and its heat dissipating properties ultimately limit the mean power that can be

*Microwave source requirements* 31

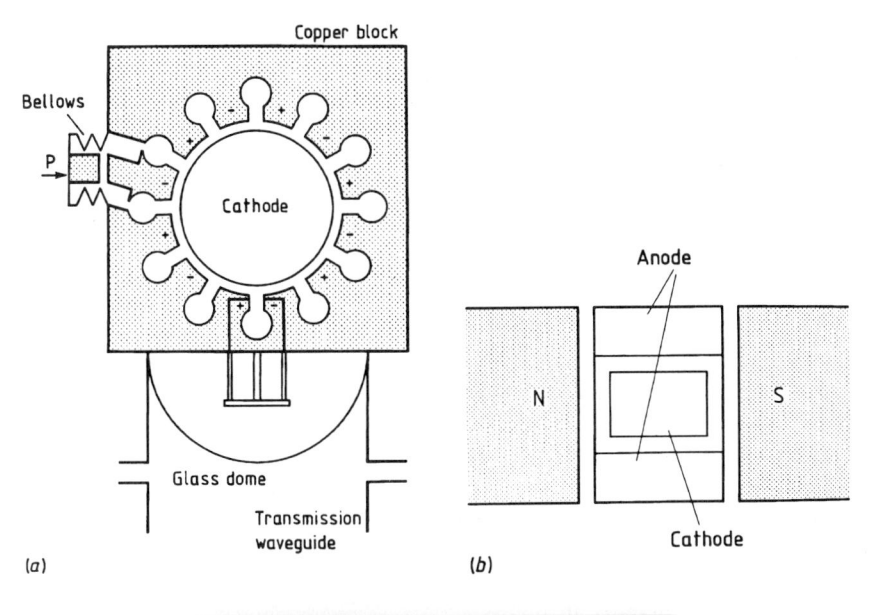

**Figure 3.4.** *(a) A diagram of a tuneable magnetron. This is placed in a magnetic field with lines of force normal to the plane of the diagram. (b) A section through the magnetron normal to that in (a), showing the poles of the magnet. (c) A photograph of a magnetron which has been opened to show the anode–cathode structure.*

generated by a given valve. Magnetrons were originally developed as power sources for radar transmitters where pulsed output was a basic requirement. As already stated in the previous chapter, the instantaneous power levels required in the accelerating waveguide corresponds to the peak electron energy levels that can be generated and this accounts for need the pulsed operation of the whole system.

Under normal operating conditions most of the emitted electrons in the magnetron eventually reach the anode but a significant proportion return to the cathode with reduced energy and raise its temperature. To maintain the cathode at constant temperature the cathode heater voltage must be reduced from its standby value as the mean magnetron power is increased. At full power the current through the cathode heater may be reduced to zero. The heater current can be regulated by measuring the mean power drawn from the HT power supply, which is directly related to the mean magnetron power. Alternatively it can be regulated to pre-programmed heating levels which take into account the mean power, related to the pulse repetition frequency, pulse voltage and current.

The arrangement for extracting microwave energy from the magnetron is illustrated schematically in figure 3.4(a). Wires from opposite sides of one of the oscillating cavities are brought through holes in the anode block to form a pair of symmetrical loops which couple directly into the transverse magnetic mode of the transmission waveguide. These output loops are under the glass dome which seals off the vacuum inside the valve.

The magnetron is a self-exciting oscillator whose frequency is determined by the mechanical shape of the resonant cavities in the anode block. The frequency can be varied slightly by moving the plunger P in figure 3.4(a) to vary the resonant frequencies of the cavities to which the cavities containing P are connected. To vary, or control the frequency requires very precise movements of P and this is done by rotating the knob of a micrometer mechanism. This is coupled to the plunger, inside the vacuum in the magnetron, via a set of bellows. It is necessary to regulate the magnetron frequency by a feedback system which controls the position of the tuning mechanism. AFC is discussed in section 3.4.

For higher-power operation, required in linear accelerators producing x-rays of energy greater than about 10 MeV, a 5 MW magnetron is required. This higher power level can be achieved by increasing the length of the anode and cathode structure shown in figure 3.4(a) in the direction normal to the plane of the diagram and supplying the magnetic field by the use of a coaxial coil. This system then allows the possibility of changing the magnetic field strength, and consequently changing the voltage at which conduction is established, which in turn allows the microwave power output to be varied. This is utilized in some linear accelerators where two or more sets of operating conditions for the 'long-anode' magnetron are used, permitting operation at peak power of between 2.5 and 5 MW. This allows the system

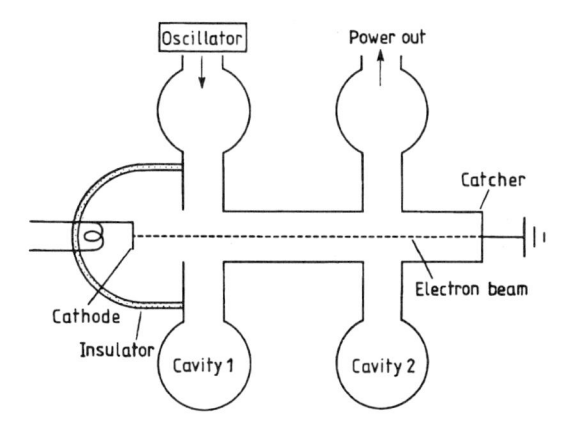

**Figure 3.5.** *An outline diagram of a klystron. Cavities 1 and 2 are referred to as the primary and secondary cavities in the text.*

to accelerate the electron beam current in the accelerating waveguide to different energies, and therefore to produce alternative x-ray qualities from the same linear accelerator.

### 3.2.2. The klystron

The major elements of a klystron are illustrated in figure 3.5. The primary resonant cavity is excited by the use of a low-powered oscillator of the required frequency. If a negative-going voltage pulse is applied to the cathode an electron beam will pass through the system as shown. In passing through the gap in the first cavity, the oscillating field will again either accelerate or decelerate the electrons. Electrons early in phase will then travel on more slowly than those passing through the cavity later in the phase. At some distance downstream the early-phase electrons will be overtaken by the later-phase electrons, so at this distance the electrons will arrive in bunches at a frequency determined by the resonant frequency of the first cavity. This frequency can be varied over a limited range by changing the frequency applied to the first cavity. If the second resonant cavity has the same frequency as the first, and is placed with its gap at the bunching position, it will be excited by the electron bunches arriving in resonance and the electron bunches will pass energy very efficiently to the oscillating field in the second cavity. The electrons will then pass through the second cavity and arrive at the water cooled electron catcher.

In practice the two resonators and the electron catcher are all at the same DC level, i.e. they are joined by metal flight tubes and it is convenient to operate them at earth potential. The system is then energized, as previously stated, by applying a negative-going pulse to the cathode. The incoming and outgoing microwave power can be handled by waveguides coupled to the primary and secondary cavities.

There is clearly a further electron bunching position downstream from the second resonant cavity, so if a further resonant cavity is placed here, yet more energy can be extracted from the electron beam, i.e. the power level can be built up yet further in the third cavity. In practice four-cavity klystrons are used.

The klystron is placed within a focusing coil which is coaxial with the electron beam. The magnetic field provided by this coil will then hold the electron beam on axis.

The arrangements for delivering power to the klystron are similar in principle to those described for the magnetron in section 3.5 and shown in figure 3.10. The multicavity klystron is operated at a pulse voltage of about 100 kV, and this is achieved by using a larger step-up ratio in the pulse transformer. It is also possible to operate the system at two different power levels by selecting different pulse voltages. From the description it can be seen that a klystron is a tuned amplifier. The frequency generated can be varied and controlled within narrow limits by varying the frequency of the exciting oscillator.

The electrons arriving at the catcher in figure 3.5 will generate x-rays, and because the electron current is large the klystron presents a major x-ray hazard. The x-ray quality (100 keV) allows the radiation to be contained by placing the whole klystron structure in a relatively thin-walled metal container.

There is a remarkable relationship between the operations of the magnetron and klystron and that of the accelerating waveguide. In the magnetron and klystron, a high current of relatively low-energy electrons is used to excite oscillations in a set of coupled cavities. The energy from these cavities is then passed to the accelerating waveguide (again, essentially a set of coupled cavities), where it is used to accelerate a very small electron current to high energy. In terms of this description the magnetron or klystron is the primary, and the accelerating waveguide the secondary, of rather a complex transformer.

## 3.3. MICROWAVE POWER TRANSMISSION

### 3.3.1. Rectangular waveguides

The major microwave components of the linear accelerator are connected by standard S band rectangular waveguides which transmit power in the $TE_{0,1}$ mode which is the only mode possible. Waveguide sections can be straight or can have bends and twists which allow reasonable freedom in the relative positioning of the components to be collected. Minimum attenuation is achieved with straight sections and care has to be taken to avoid sharp bends and twists which cause reflections of microwave power with the risk of the establishment of standing waves and ultimately electrical breakdown. In practice, bends can be tolerated if the radius is greater than 1.5 times the guide dimension in the plane of the bend. Twists must be slow: a 90° twist must be made many times longer than the wavelength. As most of the components including these connecting sections are rigid it is often desirable to include a flexible section in the circuit to allow some tolerance in assembly and to limit mechanical stress. Flexible sections can be constructed by corrugating the waveguide walls.

### 3.3.2. Transition sections and windows

Transition sections are needed to connect components of different types. For example the magnetron launches microwaves into a short section of circular waveguide that has to be connected to a rectangular transmission guide. Because of the high electric field strengths in the waveguide sections connecting the major functional components it is necessary to operate them either under vacuum, or at high gas pressure, to prevent voltage breakdown, even under normal operation conditions (i.e. when the build-up of standing waves is not significant). The vacuum and high-pressure systems are described further in chapter 4. The transition section, between the magnetron and transmission guide, needs to be gas filled, to provide gas cooling for the glass output dome of the magnetron, and is operated at high pressure to prevent sparking. It is typically operated at twice atmospheric pressure and filled with nitrogen, freon or $SF_6$. The external microwave load at the high-energy end of the accelerating waveguide also needs to be in a gas filled waveguide, the gas providing surface cooling for the dielectric load. Both gas filled sections of the microwave system are separated from the evacuated sections by waveguide windows. These are made of dielectric material, either ceramic or quartz, which is transparent to the microwave radiation.

### 3.3.3. Isolators

The isolator allows the passage of radiation from the magnetron, but presents a high impedance to reflected waves. It therefore prevents the build-up of standing waves in the transition section.

The isolator, figure 3.6(a), consists of a special waveguide containing ferrite material. Ferrite is a ceramic which is made by firing a mixture of iron and other oxides. It has the properties of a ceramic in that it has a sufficiently high resistance to permit the passage of microwaves, while it also has ferromagnetic properties. When this material is placed in a magnetic field the magnetic moments of the ferromagnetic atoms will line up with the

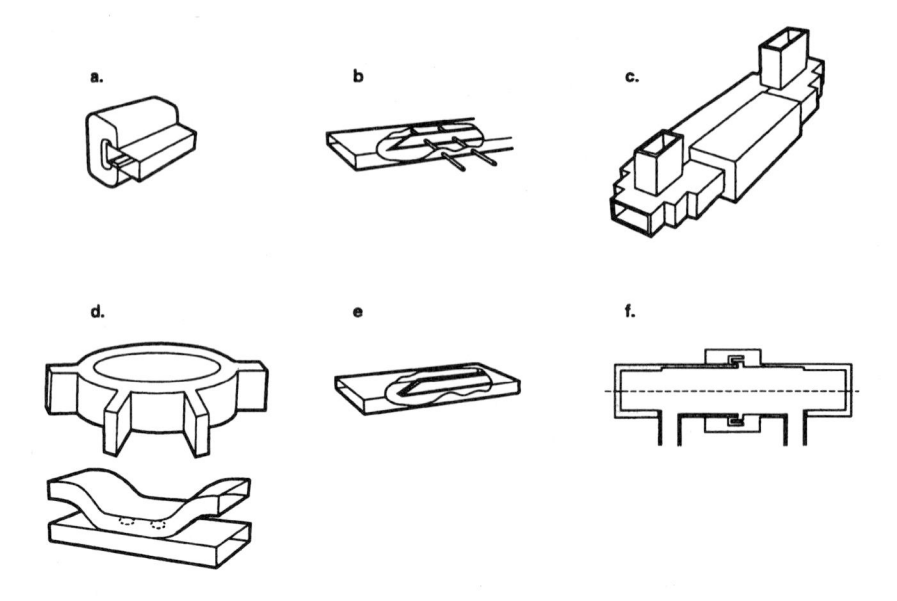

**Figure 3.6.** (a) An RF isolator, a special waveguide section containing ferrite material. (b) A phase shifter, a special waveguide section with a dielectric wedge. (c) A four-port circulator. (d) Two types of hybrid coupler, the hybrid ring and the side wall coupler. (e) An RF load, a section of waveguide containing a lossy dielectric material. (f) A rotating waveguide joint with a quarter-wavelength coaxial cavity between the sections.

field, and then precess in response to any electromagnetic disturbance, in this case the passage of microwaves. By suitable choice of ferrite material and magnetic field, provided by a permanent magnet, this precession frequency can be made equal to that of the microwave passing along the isolator. The direction of the precession, depending on the direction of the applied magnetic field, can then be either with or against the direction of the microwave signal. As a result the impedance presented to the microwave radiation is different in the forward and reverse directions.

The isolator is designed to absorb energy and when it does so it becomes hot. It is therefore part of the pressurized section of the microwave circuit and is cooled by water circulation.

### 3.3.4. Phase shifters

The introduction of a low-loss dielectric material into a section of waveguide increases the characteristic impedance and hence reduces the phase velocity. To obtain a variable phase shift a narrow slab of dielectric is mounted on ceramic rods so that it can be moved across the guide from the side, where the electric field is minimum, to the centre, where the electric field is strongest and the maximum effect is experienced. The dielectric must be carefully supported and is tapered to avoid serious reflections of microwave power. A phase shifter is illustrated in figure 3.6(b).

### 3.3.5. Four-port circulators

The four ports of the circulator shown in figure 3.6(c) are connected by two straight sections of waveguide, one of which contains a ferrite strip, which is magnetized by an external permanent magnet. As microwave power enters port 1 it is split between the two sections and recombines to leave at port 2. In this direction there is no phase change in either section. Power entering at port 2 is similarly split between the two sections but in this case the magnetized ferrite strip induces a 180° phase change in one of the sections. As a result the power leaves the device from port 3 rather than port 1. Similarly power entering at port 3 leaves port 4 and power entering at port 4 leaves at port 1.

### 3.3.6. Four-port hybrid couplers

These are devices that allow microwave power entering two input arms to be added and transmitted from two output arms, the in phase components from one arm and the out of phase components from the other. Their use in the high-power microwave circuit of a linear accelerator is to add the residual power from the waveguide to the input power from the magnetron. Ideally the phase of the fed-back power will be in phase with the source but in practice this might not be the case. Figure 3.6(d) shows two types of coupler that have been used: the hybrid ring or rat race used on early accelerators consists of four T-junctions arranged in a ring and the sidewall coupler, in use on some later machines, consists of two sections of rectangular waveguide coupled by two holes separated by one quarter wavelength.

### 3.3.7. RF loads

In general a proportion of the microwave power generated is not used and has to be dissipated as heat in a controlled way. The RF load, figure 3.6(e), is a section of waveguide containing a lossy dielectric material which is cooled either by water or, if the power dissipation is not too high, by forced air circulation over cooling fins on the metallic walls of the guide. As with the dielectric material in a phase shifter the load material is tapered to avoid power reflection which may result in standing waves elsewhere in the system.

## 3.3.8. Rotary joints

In those machines where the RF source is mounted on the stationary part of the gantry it is necessary to use a rotating joint to transfer power to the accelerator on the rotating structure.

Cylindrical waveguides can be used to form rotating joints in which the output section is allowed to rotate about the common cylindrical axis. Direct electrical connection between the two parts is undesirable as very high currents would be generated across the relatively high resistance of the junction. The solution, illustrated in figure 3.6(f), is to use a choke joint in which a gap, which corresponds to a quarter-wavelength coaxial cavity, is deliberately introduced between the sections. The metal to metal sliding connection is at the short-circuited end of the cavity, which becomes a node for RF current. As the current at the node is zero the fact that the contact resistance is relatively high is of no significance.

## 3.4. CONTROL OF MICROWAVE FREQUENCY

### 3.4.1. Magnetron driven machines

*3.4.1.1. Frequency control.* For a fixed position of the magnetron tuning plunger, P, the frequency generated depends on the temperature of the anode block, and the orientation of the magnetron in relation to the vertical. Even though the magnetron anode is water cooled its equilibrium temperature will depend on the mean power at which it is being operated. The resultant thermal expansion can cause significant changes in frequency. Also, if the magnetron is mounted on the rotating part of the linear accelerator gantry, the frequency generated will depend on the orientation of the gantry. The small but significant changes in frequency which occur are thought to be due to slight sagging of the cathode with respect to the anode block. For both of these reasons the magnetron frequency has to be continuously monitored and controlled.

As was explained in the previous chapter, it is also possible to operate the accelerating waveguide to produce different electron energies by changing the microwave frequency. It follows then from this discussion that it is necessary to run the magnetron at a selected set of frequencies, where each particular frequency can be selected and controlled with high precision. For stable operation of the accelerator system, these frequencies have to be controlled to an accuracy of a few tens of kilohertz.

The system to control the magnetron frequency is shown in outline in figure 3.7. The output from the magnetron is sampled by a probe in the transmission waveguide and is fed into two tuned cavities. These are machined out of heavy copper block, which is kept at a controlled

## Control of microwave frequency

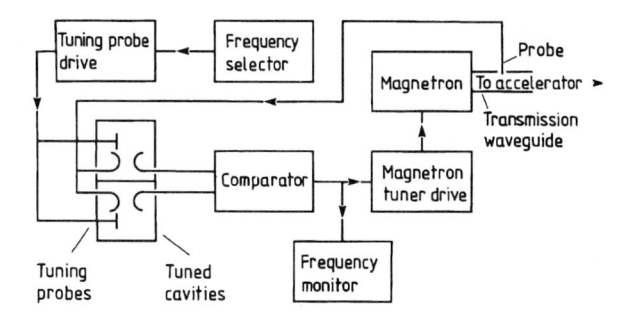

**Figure 3.7.** *A block diagram of the AFC system for a tuneable magnetron.*

temperature by the water cooling system. The cavities are tuned to slightly different frequencies, just below and just above the desired value, and the response of each cavity is sampled by the output loops shown. If the frequency is the desired value the signals from the two cavities will be equal; if the frequency is too high the high-frequency cavity will give a larger signal than the low-frequency cavity and vice versa. The input to the comparator consists of two channels, one for each cavity, each consisting of a pulse amplifier and a rectifier, and these generate signals to be compared by a circuit which controls the magnetron tuner drive. This circuit will give a null signal for equal outputs from the two tuned cavities, or, in the event of unequal signals, will give an output voltage to control a motor which turns the magnetron tuner in the direction to equalize the cavity signals.

The resonant frequencies of the two tuned cavities can be varied by changing the position of a tuning probe in each cavity. If the two tuning probes are mechanically coupled so that they always move the same amount (figure 3.7), then it is possible to determine probe positions which define a set of frequencies where the previously stated condition is satisfied, where one cavity resonates just above, and one just below, the required value. The required set of frequencies can then be controlled by selecting appropriate voltages in the frequency selector, which controls the tuning probe drive. As the selected frequencies control the energy of the accelerated electrons, the frequency selector is, in fact, one part of the electron energy selector. The system is then controlled to the selected frequency when the comparator circuit gives a null output. The output from this circuit can be used to monitor the frequency and can be calibrated in kilohertz to indicate variation of the system from the selected frequency.

The small motors driving the magnetron tuner drive and the tuning probe drive need to be servo controlled so that either system will drive to the required position at a rate proportional to the control signal. This critically damps the systems and prevents them from overshooting and hunting. This system for controlling frequency is called automatic frequency control (AFC).

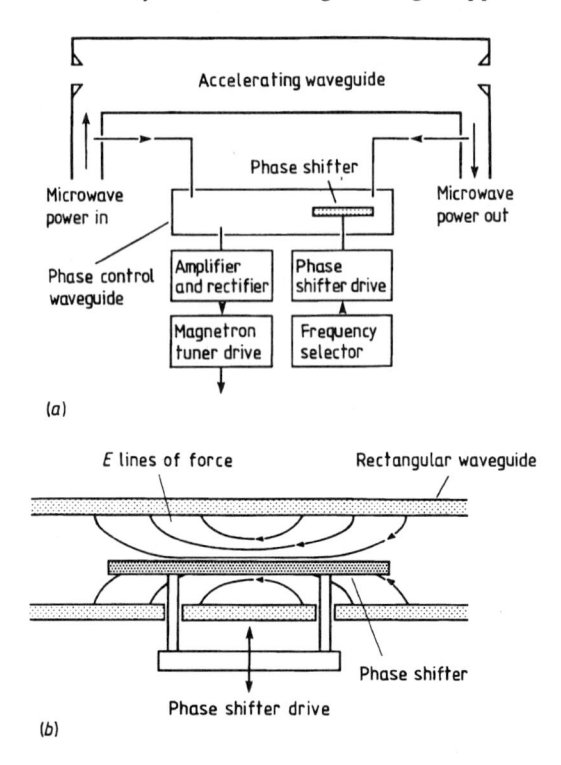

**Figure 3.8.** (a) A block diagram of the system to control the relationship between the phases at the input and output ends of the accelerating waveguide. (b) The use of a movable phase shifter.

*3.4.1.2. Phase control.* In the travelling wave accelerator, the electron energy can be varied in a controlled way by using frequency changes to control the velocity of the travelling wave. This velocity will then determine the relationship between the phases at the input and output ends of the accelerating guide. In the circuit shown in figure 3.8, these phases are compared, and a signal can be generated to adjust the magnetron frequency to bring them to the required relationship. This circuit may be used as a fine adjustment on the magnetron frequency after the AFC circuit has made its initial adjustment.

Looking at figure 3.8 in more detail, the input and outputs of the accelerating guide are sampled by probes as shown, and the signals brought into opposite ends of a length of rectangular waveguide, which contains a phase shifter section. The phase shifter is a length of dielectric material, usually ceramic, which is held parallel to the long axis of the guide. The velocity of the wave moving from right to left along the phase control waveguide in figure 3.8 will be different in the dielectric material, and therefore the phase of the wave at the left-hand end of the dielectric will be changed by its presence. If the length of dielectric is thin in relation to the inside dimensions of the waveguide (figure 3.8(b)), then the electric field passing through it will depend on its position in the waveguide, being a maximum at the centre of the guide. The effect of the dielectric thus depends on its position in the guide. By moving the dielectric in the direction shown in figure 3.8(b), the phase change between its opposite ends can be shifted. Another way of expressing this thought is to say that the effective length of the section of the guide occupied by the phase shifter can be varied as the phase shifter is moved. The phase shifter is mounted on ceramic pegs, as shown, which pass through the wall of the waveguide, and are attached to the phase shifter drive motor.

Returning to figure 3.8(a), there is then, in principle, a position for the phase shifter where the signal from the left-hand end of the phase control waveguide is in anti-phase with the signal from the right-hand end, and a null output is picked up by a probe as shown. In other words, it is possible to set the phase shifter such that a null output signal is obtained for a specified phase change between the input and output of the end of the accelerator waveguide. If the phase change is not of the required value, the output signal can be used to control the magnetron tuning drive to make a fine adjustment of the magnetron frequency until a null signal is obtained.

There is therefore a set of phase shifter settings corresponding to different frequencies which determine the energy of the accelerated electrons. These settings are made by the phase shifter drive, which is itself controlled by the frequency selector

*3.4.1.3. Coarse and fine frequency control.* As previously stated, the AFC and phase control circuits are operated as coarse and fine control. When the system is first switched on the magnetron frequency is under the control of the AFC (figure 3.7): this provides relatively coarse control. When the comparator circuit comes close to a null reading, control is handed over to the phase control circuit, which then makes the necessary fine adjustments to the magnetron frequency. Thereafter magnetron frequency control is switched between these two control systems by continuously monitoring the coarse-control signal. These functions then require a relatively complex switching arrangement between the circuits in figures 3.7 and 3.8 but these will not be detailed here.

*3.4.1.4. Programmed control.* Continuous AFC is necessary because the frequency stability of the power source is low compared with the very high $Q$ of the accelerating structure. However the frequency generated is predictable as it depends on the position of the tuner mechanism, the magnetic field, the power level and the recent thermal history of magnetron. Programmed control makes use of this information, in particular the magnetron's cooling curve, to pre-set and then adjust the tuner position to produce a frequency

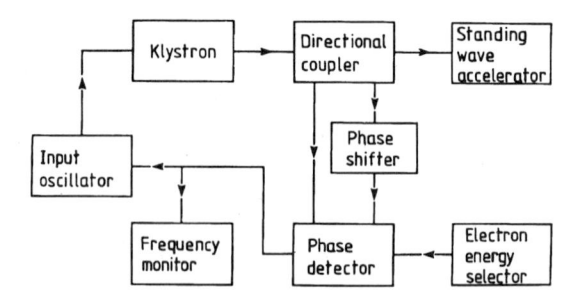

**Figure 3.9.** *A block diagram of the system to control the frequency of the klystron to hold it in resonance with the accelerating waveguide.*

which is within the capture range of the phase detector. This system has been implemented on some computer controlled machines and has the advantage that adjustment can be made simply by modifying stored control data and that multiple data sets can easily be created for the different operating conditions for which the machine will be used.

### 3.4.2. Klystron driven machines

In a standing wave accelerator, the microwave power has to be supplied at the resonant frequency of the accelerating guide. At resonance, the reflected power from the guide is exactly in phase with the incoming power, and this relationship can be utilized to control the frequency of the incoming power.

The input power to the accelerating guide is passed through a directional coupler which samples the incoming power and the reflected power. The reflected power signal is passed through a phase shifter (figure 3.9) which is set to produce a 180° phase change. At resonance, the two signals entering the phase detector will therefore be in anti-phase, and when compared will give a null output. If the incoming power is not in resonance with the standing wave accelerator, this phase detector will generate a positive or negative signal to adjust the frequency of the input oscillator towards resonance. The magnitude of the out of phase signal will depend on the power level at which the system is working. This power level may be changed to one of a set of selected values to vary the energy of the accelerated electrons, so the gain of the amplifier in the output stage of the phase detector needs to have a number of settings corresponding to the different power levels. These gain settings are controlled by the electron energy selector. The signal from the phase detector can also indicate that the signal is 'on frequency' as a null reading and can be calibrated to indicate frequency high or frequency low.

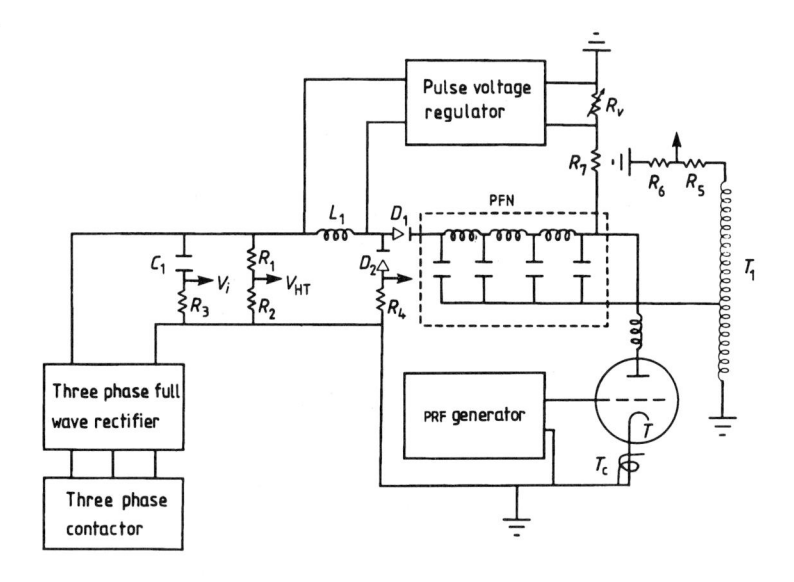

**Figure 3.10.** *The modulator circuit, which generates high-voltage pulses to energize the microwave generator.*

## 3.5. THE MODULATOR

This section gives a simplified description of the pulse modulator for a magnetron driven machine.

The prime function of the modulator circuit is to supply negative-going high-voltage pulses to the cathode of the microwave generator valve, its anode being at earth potential, although as will be seen in chapter 5 the same pulses are applied to the electron gun. The main elements in this circuit are shown in figure 3.10. The three-phase full-wave rectifier uses solid state diodes and delivers about 10 kV to the smoothing capacitor $C_1$. This voltage can be monitored at the earth end of the high-value resistance chain $R_1$, $R_2$ as $V_{HT}$. The voltage $V_i$ across the resistance $R_3$ in series with $C_1$ is proportional to the current from the high-tension power supply, and can be utilized to control an auxiliary circuit, not shown, which will open the three phase contactor in the event of an overload.

The pulse forming network (PFN), or artificial transmission line, is charged through inductance $L_1$ and diode $D_1$, while the hydrogen thyratron T is in its non-conducting state. Under these circumstances the inductance $L_1$ and the lumped capacity of the PFN act as a series resonant circuit (the total inductance in the PFN is negligible compared with $L_1$), and the voltage across the PFN will swing up to twice that from the power supply. After the voltage has reached its maximum value, it will be held by the presence of diode $D_1$. When the high-powered hydrogen thyratron is fired

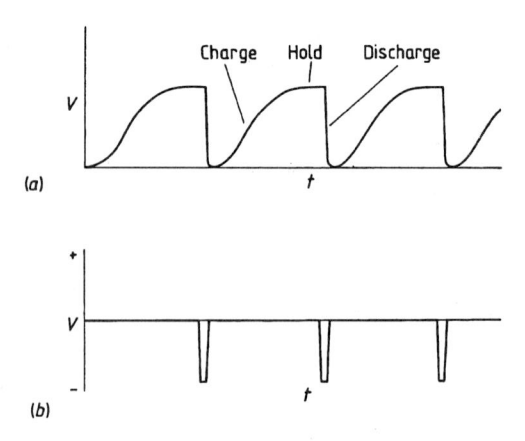

**Figure 3.11.** (a) The voltage waveform across the PFN as it is charged through diode $D_1$ and discharged via the hydrogen thyratron T in figure 3.10. (b) Negative voltage pulses at the high-voltage end of the pulse transformer $T_1$ in figure 3.10.

it will discharge the PFN, the resulting current pulse passing through the primary windings of the pulse transformer $T_1$. This is an auto-transformer with multiple parallel windings (see later), whose low-voltage terminal is at low impedance to earth and whose high-voltage terminal is connected to the cathode of the microwave generator. The length of this pulse is determined by the properties of the PFN (by the number of $LC$ 'sections'), and its voltage by that from the power supply and the turns ratio of the pulse transformer. The pulse length used is typically 3–6 $\mu$s,

The peak current through the hydrogen thyratron is of the order of 500 A and may be monitored by a current transformer $T_c$. The voltage across the PFN will follow a charge–hold–discharge cycle as shown in figure 3.11(a). The discharge stroke occurs in the time (pulse length) determined by the PFN, and the frequency is determined by that of the positive-going pulses applied to the hydrogen thyratron grid to make it conduct. The impedance of the input to the pulse transformer (under load conditions) is matched to the characteristic impedance of the PFN. If a high-voltage component on the secondary of the pulse transformer breaks down (e.g. if the microwave valve sparks), then the impedance of the primary will be suddenly reduced, and this will cause a reverse voltage pulse to travel along the PFN. The function of the 'inverse diode' $D_2$ is to conduct the resultant current to earth under these fault conditions. The voltage developed across $R_4$ in figure 3.10 is an indication of the fault in question. If the mean value of this voltage exceeds a predetermined value, this may be used to shut down the pulse repetition frequency (PRF) generator which controls the hydrogen thyratron, and thus protect the high-voltage circuits.

The positive-going signals which are applied to the grid of the hydrogen thyratron are supplied by a pulse generator whose frequency (the PRF) is usually variable. Although in principle this could be based on a free running oscillator whose frequency is continuously variable, it is usually a mains frequency based system, which operates at fixed multiples of mains frequency. The advantage of working at pulse repetition frequencies locked to the mains is that any beating effects with mains ripple on the HT power supply can be avoided. On any particular pulse the hydrogen thyratron conducts when its grid becomes positive, after which grid control is lost. Conduction stops when the anode voltage, supplied from the PFN, drops to a sufficiently low value. The impedance of the charging circuit, $L_1D_1$, is too high for any significant recharging of the PFN to occur during the discharge stroke. The pulse waveform generated at the high-tension end of the pulse transformer is shown in figure 3.11(b) and can be monitored at the earth end of the high resistance chain $R_5R_6$ in figure 3.10.

To ensure stability of the operation of the microwave generator, the voltage of the PFN needs to be regulated. This voltage is monitored from the resistor network $R_7R_v$, the signal across $R_v$ being used to control the pulse voltage. When this signal reaches a predetermined value, the output element of the pulse voltage regulator puts a low impedance in parallel with the charging inductance $L_1$. When this happens the resonant charging of the PFN suddenly stops, and the voltage on the PFN is held at the value attained at that instant by the high inverse impedance of the diode $D_1$, until the hydrogen thyratron fires on the next cycle. The value of the stabilized voltage on the PFN may be set by the use of the variable resistor $R_v$. The process described in this paragraph is sometimes referred to as '$Q$ spoiling'.

The arrangements for supplying power to the magnetron are shown in more detail in figure 3.12. The pulse transformer in the modulator circuit is an auto-transformer with a multiple parallel winding and supplies about 50 kV pulses to the cathode. The self-capacity between these windings is sufficient to hold them at the same pulse voltage. At the earth end, two windings are connected to the output of the magnetron heater supply as shown. This is at a low impedance to earth and so the lower end of the windings can be regarded as a 'pulse earth'.

Summarizing this description of the modulator circuit, we can say that its function is to supply voltage pulses to the microwave generator, the PRF being determined by the pulse generator which controls the hydrogen thyratron grid. The peak voltage, the peak power and the mean power required from this circuit are ultimately determined by the working requirements of the microwave generator. Within this limit the dose rate from the accelerating guide may be regulated by controlling the PRF.

*3.5.1. Waveforms*

Only a few key waveforms will be considered, in so far as they illustrate the operation of the system as a whole. Clearly a large number of different

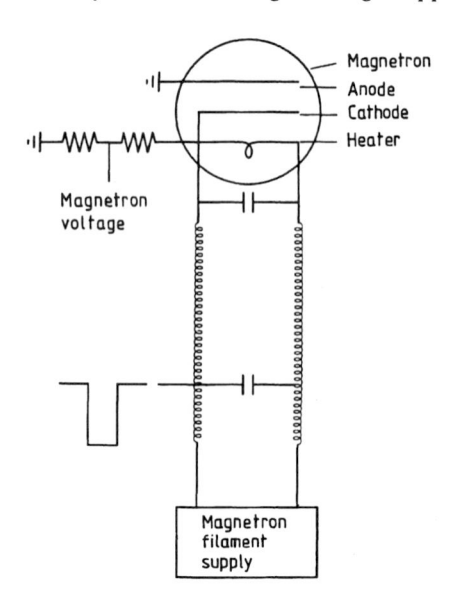

**Figure 3.12.** Parallel windings on the pulse transformer $T_1$ in figure 3.10, used to supply the heater current for the magnetron.

waveforms have to be monitored from time to time for setting up or fault finding purposes but they will not be further discussed here.

The time sequence in accelerator operation is shown in figure 3.13. Note that the time scales for figures 3.11 and 3.13(a)–(d) are very different. Figure 3.13(a) shows the negative-going voltage pulses which energize the magnetron and electron gun. These are about 3–6 $\mu$s long, varying between different makes and models but constant for a particular machine, and separated by a time determined by the PRF. For 300 pps (pulses per second), which is a fairly common value, the time scale is as shown in figure 3.13(a).

The magnetron voltage pulses are as shown in figure 3.13(b), which at first sight is very different from the ideal square pulse. The leading edge is quite sharp, and the reason for the trailing rear edge is that the impedance of the magnetron increases sharply as the voltage starts to fall. In other words, as the voltage on the anode of the magnetron starts to fall, the electrons are prevented by the magnetic field from reaching the anode, so that remaining space charge can only leak away slowly. Fault conditions, such as a spark in the magnetron itself, or in any of the high-power microwave components, will break up the flat top on the pulse.

If the electrons from the accelerator guide are being used for x-ray production, or passing through a thin window to give an electron beam in air, the radiation pulse will be as shown in figure 3.13(c). This has quite sharp edges because microwave generation by the magnetron decreases rapidly

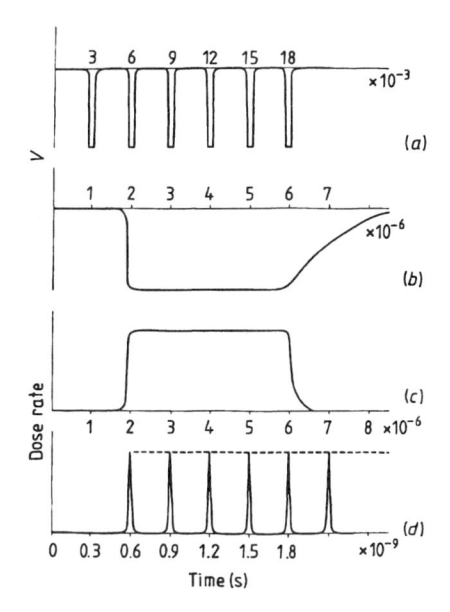

**Figure 3.13.** *Various waveforms relevant to the operation of an accelerator. (a) Voltage pulses from the modulator at the PRF. (b) An individual magnetron voltage pulse. (c) The radiation output pulse, at the same frequency as (a) and the same pulse width as (b). (d) Radiation pulses at the microwave frequency corresponding to individual electron bunches in the accelerator.*

as soon as the anode voltage starts to fall, so no electrons are accelerated in the waveguide during the period of the trailing edge of the magnetron pulse. The radiation pulse shown in figure 3.13(c) is the envelope of a series of much shorter pulses as shown in figure 3.13(d), each corresponding to the arrival of a bunch of electrons formed in the buncher section of the accelerating guide. These pulses are then separated in time by one cycle of the microwave radiation, i.e. by about 0.3 ns for 3000 kHz.

The only practical consequence of the pulsed nature of the radiation output in relation to machine design arises from the fact that the instantaneous radiation dose rate during the pulse is very high compared with the mean dose rate. For 4 $\mu$s pulses at 300 pps, the dose rate during a pulse is $\frac{5}{6} \times 1000$ times the mean dose rate. The monitors in the treatment head have to be designed to cope with these very high dose rates.

# CHAPTER 4

# THE VACUUM, COOLING AND ANCILLARY SYSTEMS

Because of the high electric field strengths in the waveguide sections connecting the major functional components it is necessary to operate them either under vacuum, or at high gas pressure, to prevent voltage breakdown under normal operation conditions (i.e. when the build-up of standing waves is not significant). Additionally all parts of the electron beam transport system, including the electron gun, the accelerating waveguide, and the flight tube and bending chamber, if used, have to be operated under vacuum to prevent scattering of the electron beams.

## 4.1. THE VACUUM SYSTEM

Those elements which form the vacuum system are shaded in figure 4.1. These include the transmission waveguide, the electron gun, the accelerating waveguide and the beam transport system as well as the roughing pump and removable flexible connection needed to establish the high vacuum. The evacuated sections of the microwave system are separated from gas filled sections by waveguide windows: these have been described in section 3.3.2.

The minimum vacuum condition required for operation of the system is determined by the fact that electrons being accelerated should not be deflected by collisions with gas atoms. In other words, the mean free path between collisions with gas atoms needs to be long compared to the length of the accelerating waveguide, and the subsequent flight tube. This length is typically 1–3 m. The maximum gas pressure to meet this condition is about $1.3 \times 10^{-3}$ Pa or $10^{-5}$ torr (1 torr = 1 mm Hg = 133.3 Pa: the pascal is the SI unit of pressure). Some early linear accelerators for medical use operated at this pressure, being pumped by oil diffusion pumps without using a cold trap. To operate a clean system which gives good cathode life for the electron gun, it is advisable to operate at pressures at or below $1.5 \times 10^{-5}$ Pa. Present-day machines work at this level.

*The vacuum system* 49

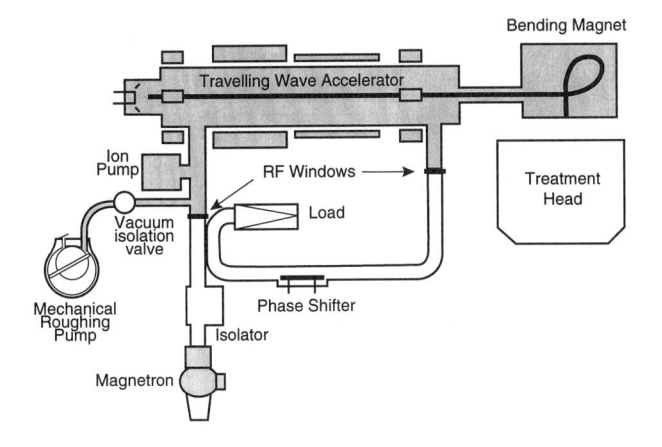

**Figure 4.1.** *The vacuum system showing parts of the accelerator which are continuously evacuated (shaded) and the arrangements for establishing a high vacuum.*

## 4.1.1. Ion pumps

Current technology uses ion pumps to maintain the pressure in the vacuum system. The ion pump operates as a cold cathode discharge tube. By running the gas discharge in a magnetic field, electrons travelling between cathode and anode follow a spiral shaped path, so increasing the number of electron–gas molecule collisions, thus making it possible to run a self-sustaining discharge at very low pressures. The magnetic field is provided by a permanent magnet and the ion pump body incorporating the anode and cathode fits between its poles. Ionized gas atoms are accelerated towards the cathode, by the applied anode–cathode voltage, where they produce secondary electrons, which keep the discharge going. The ions striking the cathode also have enough energy to eject or sputter atoms from its surface and these sputtered atoms are deposited on the walls of the pump. Gas atoms striking the freshly sputtered layer are adsorbed and are subsequently buried in the next layer of sputtered material. To facilitate the action described, the cathode should be made of a material such as titanium which has a high sputtering coefficient. The pumping action is then provided by trapping gas atoms on a continuously renewed surface of sputtered material (for further information concerning vacuum technology see Carpenter 1983).

The gas discharge current is clearly dependent on the gas pressure, and so a meter indicating this current can be calibrated as a pressure gauge. The working range of ion pumps is $10^{-1}$–$10^{-6}$ Pa. The control unit for the ion pump consists of an HT supply (3–5 kV) to operate the gas discharge. A control signal proportional to the current can be used within other parts of the accelerator's control circuitry to provide interlocks which will prevent operation of the modulator and the electron gun filament supply if the pressure rises above a pre-determined value. Linear accelerators may be fitted with one or more ion pumps depending on the evacuated volume with a pumping speed of typically 25 l s$^{-1}$ necessary for overcoming very small leaks and the unavoidable outgassing of new and replacement vacuum components.

### 4.1.2. Roughing pumps and operation of the vacuum system

Since the upper working limit for the ion pump is $10^{-1}$ Pa, it is necessary to have a valve and coupling arrangement so that initial pump-down from atmospheric pressure can be carried out with a rotary vacuum pump. It is usually more convenient to arrange for the mechanical pump to be temporarily coupled for this purpose, and it is also necessary to include an additional vacuum gauge in the mechanical pump system so that pressure can be indicated during this stage. A Pirani hot-wire gauge is usually used. As the ultimate pressure of a rotary vacuum pump is close to the maximum pressure at which the ion pump can start operating other types of pump with lower operating pressures have been used. These include oil diffusion pumps, which must be fitted with a cold trap to avoid the danger of oil vapour contaminating the accelerator, and turbo-molecular pumps, which can reach very low pressures by virtue of the turbine blades moving much faster than the molecular motion of the gas that they are displacing. The connection of the roughing pump to the accelerator's vacuum system requires careful consideration as the pumping speed is critically dependent on the cross sectional area of the interconnecting lines and the aperture through the isolation valves when they are open.

Most of the components of the vacuum system are connected using compressible soft metal gaskets which give less leakage than neoprene O-ring gaskets. Materials used for the gaskets may be pure copper or indium. The necessary pressure in the gasket seal may be achieved by using a knife edge and flat plate construction to bear on the soft metal. An O-ring seal may be used on the electron gun mounting, as this needs to be readily demountable and is subject to thermal cycling. The tolerable leakage on the system is such that it is advisable to use a mass spectrometer leak detector for leak hunting, using helium as the probe gas.

The most common reason for letting the vacuum system fill with gas to atmospheric pressure is to change an electron gun cathode. On average this should not be required more than once a year and is never required on some small accelerators which are 'sealed for life'. The actual pumping out and degassing procedure is very dependent on the cleanliness of the system: even grease from finger marks can be a significant source of troublesome vapours. Vacuum components should be handled with gloved hands and any surface cleaning carried out with an approved solvent. In order to avoid problems with water vapour, which can condense on the cold inner surface, the system should be filled with pure dry nitrogen, rather than air, before dismantling the system.

The second most vulnerable components in the vacuum system are the x-ray target and thin window at the end of the drift tube. These undergo regular heating and cooling cycles during normal machine operation and so may eventually develop small cracks. This is not likely to happen more than a few times during the lifetime of the system, if the flight path of accelerated electron beam is being adequately controlled.

The procedure in pumping out the system from atmospheric pressure is not trivial as gas will have been absorbed into the internal surfaces of the vacuum components, even when great care has been taken with cleanliness.

When the roughing pump has achieved its ultimate pressure it has to be sealed off and pumping switched to the ion pump, which will eventually sustain the operating vacuum pressure. At this stage the pressure should be $10^{-1}$–$10^{-2}$ Pa. The valve in the mechanical pump port is closed and when the ion pump is switched on the pump current will be high, and initially the gas pressure will start to fall. If the pump is allowed to run continuously under these conditions, the high current will cause it to heat, eventually to the point where the sputtered metal will start to evolve gas rather than trap it. If this occurs it is necessary to switch off the pump and allow it to cool. The gas pressure can then be brought down by a succession of pumping and cooling cycles of the ion pump, until eventually the pump current reaches a sufficiently low value to allow continuous operation without significant heating. When the gas pressure falls below about $10^{-3}$ Pa, the vacuum interlock will allow heating of the electron gun cathode which will have absorbed some gas onto its surface. Degassing of the cathode is carried out by slowly increasing its temperature to its normal operating level and if necessary going through a sequence of heating–cooling cycles of the gun cathode, until steady pressure conditions are obtained with the cathode heater continuously energized. If the microwave system is now energized, the waveguide system will degas as energy is fed into it, so again a sequence of degassing and cooling cycles is necessary. Finally, when the whole system is energized, i.e. when an electron beam is produced and accelerated, further degassing will occur, particularly at the x-ray target or thin-window end of the drift tube. Stable operation will be achieved following these final cycles of degassing and cooling.

It can be seen that pumping out of the system is a long process, requiring much patience from the operator. The actual time required for the whole operation will depend on the previous history of the system to be evacuated, and at best will be several hours. After the vacuum system has stabilized at a pressure which allows useful operation of the accelerator it will then degas for several days, finally settling down to a pressure of less than $10^{-5}$ Pa.

## 4.2. HIGH-PRESSURE GAS SYSTEMS

The transition section, between the magnetron and transmission guide, needs to be gas filled, to provide gas cooling for the glass output dome of the magnetron, and is operated at high pressure to prevent sparking. It is typically operated at twice atmospheric pressure and filled with nitrogen, freon or $SF_6$. The external microwave load at the high-energy end of the accelerating waveguide also needs to be in a gas filled waveguide, the gas providing surface cooling for the dielectric load. Both gas filled sections of the microwave system are separated from the evacuated sections by the waveguide windows which have been described in chapter 3. The high-pressure system is usually fitted with a gauge and pressure operated switch which provide an interlock, inhibiting the operation of the modulator if the pressure falls below a pre-set level.

## 4.3. THE WATER COOLING SYSTEM

Many of the parts of the linear accelerator which need to be water cooled have been mentioned in the previous chapters: these include the microwave generator, the accelerating waveguide structure, RF isolators and loads, tuned cavities used for frequency control and some of the high-power transformers in the pulse modulator circuit. Other parts including beam focusing and steering coils, and the x-ray target, will be discussed later. For some of the components cooling is necessary in order to maintain precise temperature control for stability of operation. For example the resonant frequencies of the tuned cavities used for frequency control are affected by thermal expansion, so precise temperature control of these components is a prerequisite of the AFC system. For other components, such as the x-ray target, temperature control is less critical and is needed only to prevent overheating and functional failure. Although there are different requirements for cooling it is convenient to use the same cooling system for all these purposes and this requires that each component to be cooled must be supplied with water at a fixed flow rate and temperature.

A typical simple cooling system is shown in figure 4.2. Water is circulated from a reservoir by a pump feeding a manifold via a filter which removes any fine particles which could become lodged in the smaller bores of the cooled components. As the amount of power dissipated in each component is quite different, each of the separate circuits has a flow control valve and a flow detector shown here as a flow switch (or flow operated relay). These flow switches, a part of the accelerator's complex interlock system, prevent operation of the accelerator if one of these items is not cooled. Pressure gauges are fitted each side of the filter, the first as an indication that an adequate pressure is being maintained and the second that the pressure drop

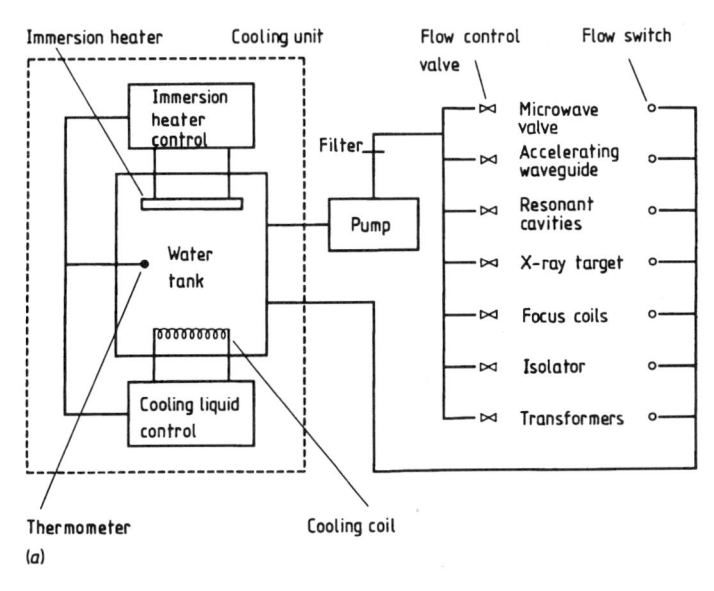

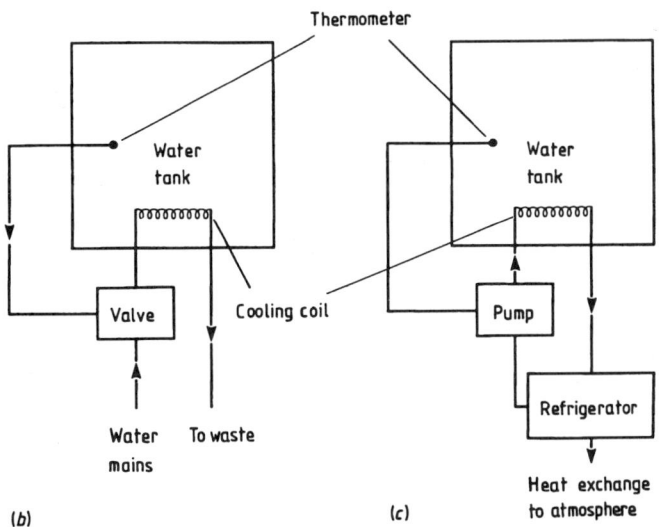

**Figure 4.2.** Schematic diagrams of a water cooling system: (a) the overall system; (b) a detail showing a water to water heat exchanger using mains water; (c) a detail showing a water to air heat exchanger via a refrigerator.

across the filter (an indication that it is not blocked) is acceptable. The temperature of the water tank is held constant to 1 °C by cooling coils, which are activated by a thermostatic control system. It may seem paradoxical for the cooling system to be fitted with an immersion heater but there are good reasons for this. At the most mundane level if the water reservoir is accommodated in an unheated space the heater can be used for frost protection, to prevent the water from freezing. It can also be used to bring the water temperature up to normal operating conditions when the accelerator is prepared for use each day. Finally the heater can be used to maintain the temperature during periods when the high-power devices are not operating, particularly in systems that have been designed to operate at temperatures significantly above ambient temperature. Operation at high temperature, or at a temperature significantly above the refrigerant temperature, allows for more efficient heat exchange and therefore more responsive temperature control than is possible with lower temperature differentials.

Cooling of the reservoir can be 'water to air' by the use of a refrigerant which then exchanges heat with the atmosphere or 'water to water', by the running of cooling water from the water mains to waste (a practice not encouraged by most water authorities), or water to water in which the energy is recycled, perhaps as an input to the hospital's domestic hot-water supply.

As the reservoir is large, the pump is noisy and the heat exchange (for water to air) needs a well ventilated space, water coolers are often sited remotely from the treatment room. The long lengths of water pipes have to be designed to minimize pressure losses. Although the use of a large reservoir has some advantages in maintaining temperature stability there are other advantages to be gained from a smaller-volume system which is sealed and can be pressurized for the primary cooling circuit. The purpose of pressuring the system is to prevent cavitation that would otherwise occur in high-velocity flow through small-bore pipes. Secondly, the quality of the circulating water is an important consideration, as in some areas the local mains water supplies are not sufficiently clear of particulate material and dissolved salts.

The use of distilled water to charge the reservoir prevents chemical, and even biological, activity which can result in corrosion and/or blockage of pipes and water ways in cooled components. It should be noted that demineralized water is not a good alternative as it can actively leach materials from the surface of components until an equilibrium is reached.

A second reason for requiring the use of distilled water arises if some of the cooled components have to be electrically isolated from earth. In these circumstances a small-volume system becomes a practical necessity, to avoid the need to provide vast quantities of distilled water. With a small-volume primary cooling system it is often necessary to use a more conventional system to provide secondary cooling and eventually discharge the heat to the atmosphere. The water to water heat exchanger between the primary and secondary systems can then be mounted on the accelerator's gantry, so that the volume of the primary circuit is minimized.

# CHAPTER 5

# THE ELECTRON BEAM (ITS PRODUCTION AND TRANSPORT)

The electron beam originates at the cathode of the electron gun, gains energy in the accelerating waveguide and passes through a beam transport system on to the x-ray target, or through a thin foil window in the treatment head which is described in the next chapter. This chapter describes the systems to guide the beam to the treatment head ensuring that it arrives precisely in the required position and direction. Magnetic and electric fields are used to control the electron beam so, before describing the functional components required for beam production and transport, a brief review of electron beam optics is necessary.

## 5.1. BEAM OPTICS

By definition, an electric field, $E$, is described by the force (a vector quantity requiring specification of magnitude and direction) on a unit charge. So, in a vacuum, a charged particle such as an electron will be accelerated in an electric field, by a force, $F_E$, in the direction of that field given by

$$F_E = -eE. \tag{5.1}$$

A magnetic field, $B$, normal to the direction of motion of a charged particle with a speed $V$ (a scalar quantity) exerts a force which is mutually perpendicular to the direction of motion and the direction of the field. If the direction of motion is not normal to the field, say at an arbitrary angle $\theta$, then the force, $F_B$, will be proportional to the component of the field that is normal to the direction of motion.

$$F_B = BeV \sin \theta. \tag{5.2}$$

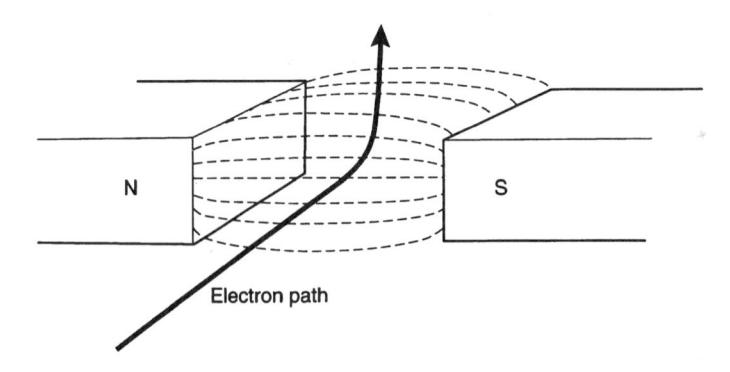

**Figure 5.1.** *The path of an electron through the dipole field.*

### 5.1.1. Magnetic dipoles

The magnetic field from a magnetic dipole is shown in figure 5.1. It can be formed between the north and south poles of permanent magnets or between the coils of a pair of electomagnets. An electron travelling in the plane midway between the pole pieces will be deflected as shown. Calculation of the complete trajectory of the electron is complicated because the magnetic field is not uniform. However in the region of uniform field between the poles the electrons follow a circular path of radius $\rho$ such that the centripetal force $mV^2/\rho$ balances the magnetic force given above, so the radius of the trajectory is given by

$$\rho = \frac{mV}{eB}. \qquad (5.3)$$

In general when we are considering electrons travelling through linear accelerators we are more interested in the electron's energy, $T$, rather than its velocity. Using the relativistic relationship between velocity and energy this equation can be transformed to

$$\rho = \frac{0.1704}{B} \left\{ \left[ \left( \frac{T}{m_0 c^2} \right) + 1 \right]^2 - 1 \right\}^{1/2} \qquad (5.4)$$

where $\rho$ is measured in centimetres, $T$ in millions of electron volts and $B$ in tesla. Karzmark *et al* (1993) simplify this equation for electrons travelling near the speed of light to

$$\rho = \frac{T + 0.511}{3B}. \qquad (5.5)$$

The complete trajectory will clearly not be perfectly circular as the field at the periphery is not uniform but at any point it will form a circular arc whose radius depends on the local magnetic field.

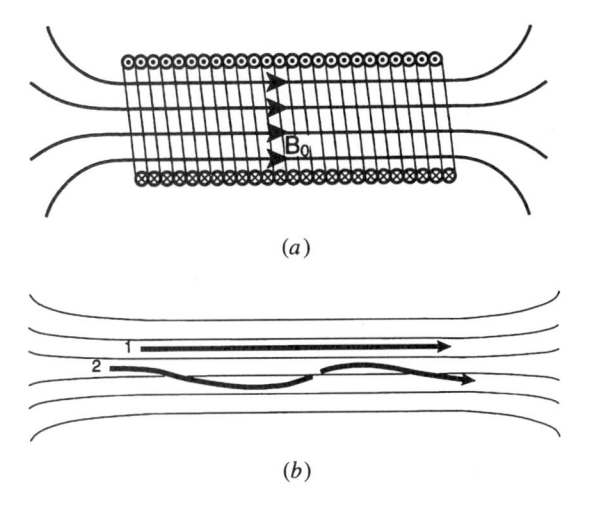

**Figure 5.2.** (a) Field lines for a solenoid. (b) Paths of electrons through the solenoid field.

Although the example shown here is for an electron which is deflected as it crosses the magnetic field of a dipole, electrons that originate in a uniform field which is strong enough go into complete circular orbits. A special case of interest is the magnetron where there is also a radial electric field in the plane normal to the magnetic field. Here the electrons travel in a complex cycloidal motion which, by interaction with fields from the resonant cavities, gives rise to oscillations at microwave frequencies.

It can be seen from equation (5.5) that as a beam of electrons pass through the field of a dipole magnet the deflection that they suffer is dependent on their energy, so a dipole magnet acts as an energy spectrometer. This and other second-order effects will be discussed later in section 5.3.3, describing the application of dipole magnets to beam bending.

## 5.1.2. Solenoids

The magnetic field created by a solenoid, a long coil energized by a direct current, is shown in figure 5.2(a). Inside the coil the field is uniform and axial but at the ends the field lines diverge. In figure 5.2(b) electron 1, travelling inside the solenoid parallel to the axis and the field lines is unaffected, as $\sin \theta$ in equation (5.2) is equal to zero. However electron 2, whose direction is not parallel to the field lines, is subjected to a force which converts its trajectory onto a helical path with an axis parallel to the field. The pitch of the helix is related to the axial velocity and the rotational frequency (the cyclotron frequency), which is proportional to the magnetic field, $f/B = 2.8 \times 10^{10}$ Hz T$^{-1}$. Figure 5.2(b) shows many turns to illustrate the process and is representative of slowly moving electrons in a strong

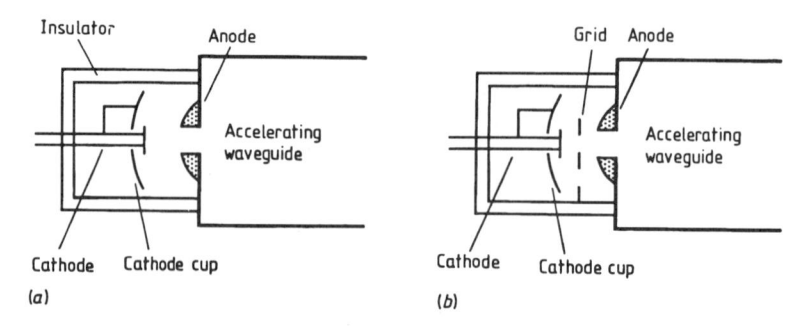

**Figure 5.3.** *(a) A diode electron gun. (b) A triode electron gun.*

solenoid field. The effect of this conversion of radial to rotational motion is to limit the divergence of electrons whose direction are not coaxial with the field on entry to the solenoid field, or whose direction is changed by other forces (e.g. radial electrostatic forces) within the region of the solenoid field.

Although the purpose of a solenoid is to create a uniform axial, the fringe fields at each end are unavoidable. In these regions electrons travelling parallel to, but displaced from, the axis will be deflected so that when they reach the uniform field they will, as electron 2 in figure 5.2, be forced onto a helical path as described. This might be seen as an unfortunate effect but a complementary effect occurs as the helically rotating electron leaves the solenoid. That is the helical path is converted back to one with either displacement or divergence from the axis. Karzmark *et al* (1993) make the interesting observation that the helices formed by an electron entering a solenoid with displacement but no divergence are tangential to the axis. This property can be exploited in the use of solenoid fields to produce convergence of the electron beam in a linear accelerator.

## 5.2. ELECTRON GUNS

Diode and triode type electron guns are used by different manufacturers, as are a wide variety of types of thermionic cathode.

### 5.2.1. *Diode guns*

The accelerator's waveguide system is normally operated at DC earth, and the gun cathode has to be supported on a glass or ceramic structure. This structure acts as an insulator and as part of the vacuum envelope for the whole system and is shown in section in figure 5.3(a). Electrons thermionically emitted from the cathode are electrostatically focused into the accelerating waveguide by the use of the curved cathode cup and anode, when the cathode is pulsed negative relative to earth potential. The electron

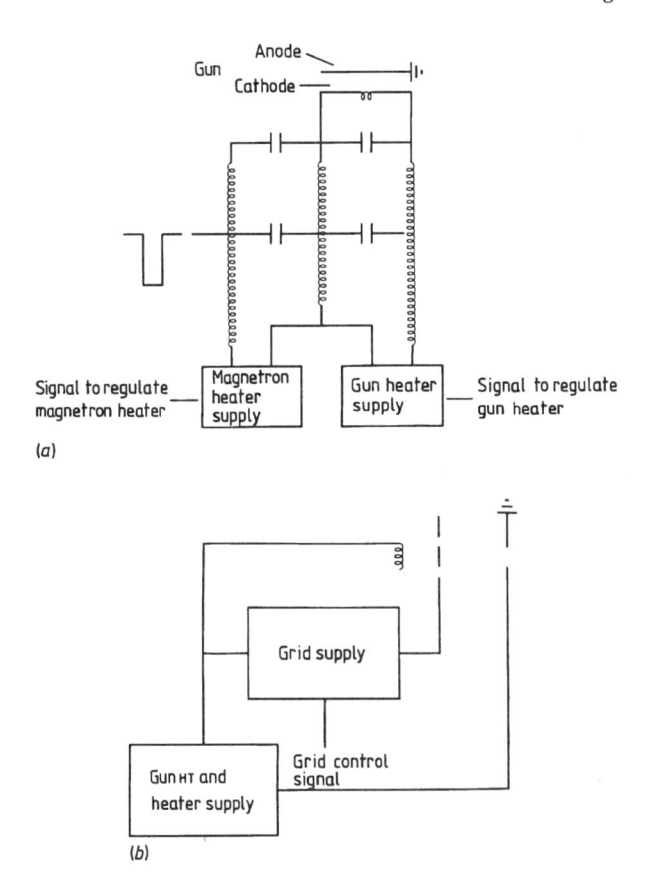

**Figure 5.4.** (a) *The arrangement for supplying cathode heater current to the diode electron gun.* (b) *The circuit to supply cathode heater current, negative cathode voltage and grid voltage to the triode electron gun.*

beam is formed by the emission of electrons through a small hole in the anode.

The arrangements for energizing the diode gun are shown in figure 5.4(a). The pulse transformer, which supplies high-voltage, negative-going pulses to the electron gun and magnetron in travelling wave machines, is drawn here in more detail than in the diagram of the modulator, figure 3.10. This transformer has multiple parallel windings, which allows the cathode heater currents for both the magnetron and the electron gun to be supplied and regulated by circuits that operate near to earth potential. The arrangements for controlling the magnetron heater have already been discussed in chapter 3. The negative-going HT pulses to the electron gun and the magnetron cathodes do not have to be at the same voltage with this system, although it is very convenient if this is the case. If the accelerator is designed to accept lower-energy electrons from the gun then the gun pulse is derived from the magnetron pulse via a simple potential divider.

Cathodes may be directly or indirectly heated. The simplest directly heated cathode is a flat spiral of tungsten wire, placed as in figure 5.3(a). The electron current into the guide can then be regulated by controlling the cathode filament current and hence its temperature. During x-ray production, when beam currents are high enough to cause beam loading, the maximum energy to which electrons are accelerated in the waveguide can be controlled by regulating the beam current. During electron beam production, when beam currents are so low that beam loading is not significant, the dose rate can be regulated by controlling the beam current. Both these control mechanisms can be achieved with a directly heated cathode in a diode gun operating in the emission limited region of the diode characteristic curve. Because the tungsten spiral has a small heat capacity and rapid heat loss occurs by visible and infra-red radiation, when it is operated as a bright emitter the electron beam current can be changed very rapidly. Fine control of the filament current can therefore be used to stabilize energy or dose rate in response to control signals related to these parameters. More details of the control mechanism are given in chapter 7.

The major change in guide current, by a factor of about a hundred, between operation of the system for x-ray therapy and for electron beam therapy is also effected by changing the gun cathode heater current.

It is possible to use an indirectly heated oxide coated cathode or other composite cathode as the electron source in a diode gun. In this case the indirectly heated structure is pill box shaped with its flat circular emitting surface taking the place of the circular spiral of a directly heated cathode. This system does not allow quick changes of electron emission because of the larger thermal capacity of its indirectly heated surface. On the other hand, it is possible to achieve sufficiently high electron emission to stabilize the guide electron current at a single value for operating the gun in the 'space charge limited' state where the beam current is determined by the voltage applied to the gun and the aperture in the anode. As the beam current for a gun operated in the space charge rather than emission limited region is inherently more stable, there is less need for an eleborate control system. However the large change in current necessary between x-ray therapy and electron therapy is more difficult to accommodate. One solution has been the use of a different anode aperture for each mode, but this requires a mechanical arrangement for changing apertures without compromising the vacuum system.

The choice of directly or indirectly heated cathode depends mainly on the need, or otherwise, for quick changes in electron beam current to stabilize the energy of the electron current.

The directly heated bright emitter has a limited life depending on its operating temperature, caused by evaporation of the tungsten, while in theory the indirectly heated composite cathodes have a much longer life. In practice the useful lives of both types are dependent on operating in a good vacuum system. Some linear accelerators require regular replacement of the electron gun cathode on site. The total time required for a gun replacement is dependent mainly on the time required to degas the new cathode, and this is a relatively rapid process with the tungsten spiral. With the oxide coated and other composite cathodes, a new cathode has to go through a degassing and 'forming' procedure during which the cathode surface is converted to the chemical form with the required low work function for high electron emission. This process takes many hours and so replacing a cathode can be a 24 h exercise. In terms of annual down time for cathode replacement, the difference between bright emitters and composite cathode is marginal.

At least one manufacturer supplies accelerators in which the electron gun and accelerating waveguide assembly are an integral structure which is delivered already evacuated. The whole assembly has then to be changed in the event of a cathode failure but in this case the cathode is engineered to a high standard and should have a life of between 5 and 10 years.

### 5.2.2. Triode guns

A triode gun (figure 5.3(b)) uses grid control of the electron beam current through the assembly, and therefore can use an indirectly heated cathode and still give very rapid regulation of the electron beam current.

The arrangements for powering and regulating the gun are illustrated in figure 5.4(b). The accelerating waveguide is at DC earth and the cathode is held at a voltage negative to earth, typically $-20$ kV, which is determined by the required initial electron energy. The grid is normally held sufficiently negative to the cathode to cut off the current to the anode, and thus the timing and magnitude of the current pulses injected into the accelerating guide are controlled by voltage pulses applied to the grid. These voltage pulses must of course be synchronized with those applied to the microwave generator. With this system it is possible to have a phase change between the microwave and the electron pulses, whereas with the diode gun the microwave and electron pulses have to be exactly in phase, the advantage being that a slight delay before injecting electrons allows the accelerator to 'fill' with microwaves and ensures that all the electrons will be accelerated to the same energy.

The grid supply and control circuit of figure 5.4(b) has to be negative to earth at the cathode voltage and so has to be supplied through suitable isolation transformers. The energy of the electrons injected into the accelerating waveguide is determined by the negative voltage applied to the gun cathode. As with the diode gun the beam current, in the form of the electron pulses, from the triode gun will also have an effect on the energy of the accelerated electrons because of beam loading, the only differences between the two systems being the precise way in which the beam current is controlled and the abilty to vary the timing and length of the pulses from the triode gun.

## 5.3. BEAM TRANSPORT

The beam transport system comprises three distinct elements, each of which uses magnetic fields to control the path of the electron beam as it passes through the accelerating waveguide and is delivered to the treatment head. The electron beam has to be 'steered' through the narrow bore of the waveguide and kept, as near as is possible to the axis. It has to be 'focused' to prevent divergence and maintain the beam with a small cross sectional area, and finally, unless the treatment head is in line with the accelerator, the beam has to be 'bent' so that it strikes a target or can emerge from the vacuum system through a thin window.

### 5.3.1. Steering coils

Electrons passing through the accelerating guide, under the influence of the microwave fields will not travel exactly along the central axis because of minor imperfections in the gun–accelerator structures, and because of the effects of external magnetic fields (the Earth's field and that due to adjacent steel items in the supporting structure, and even the building). Also, since the whole system may be rotated, the relationship between these external fields and the position of the accelerating guide is variable. The deflection of an electron with an average energy of 2.5 MeV travelling along a 1.5 m waveguide in the Earth's field of 0.5 T is typically a few millimetres; clearly the precise deflection depends on the orientation of the accelerator.

As a result of these effects the electron beam has to be actively steered through the system and this can be done by the use of two orthogonal dipoles formed by pairs of beam steering coils arranged as in figure 5.5. In broad terms the coils at the entrance to the accelerator, the gun end, steer the electron beam to the correct position immediately after it has been injected, while those at the other, high-energy, end can be used to control the direction of the beam as it leaves the accelerator, so the coils at the gun end are needed to correct for geometric misalignments in the electron gun and those at the high-energy end are needed to correct for deflection caused during acceleration by geomagnetic and other external influences. Because of the relationship of these functions to the position of the accelerator, the currents in the beam steering coils need, in principle, to be dynamically controlled. In practice the current in the gun end coils can be preset to a constant value, and the only additional requirement is that their power supplies need to be well stabilized. Failure to set, or control, the gun end

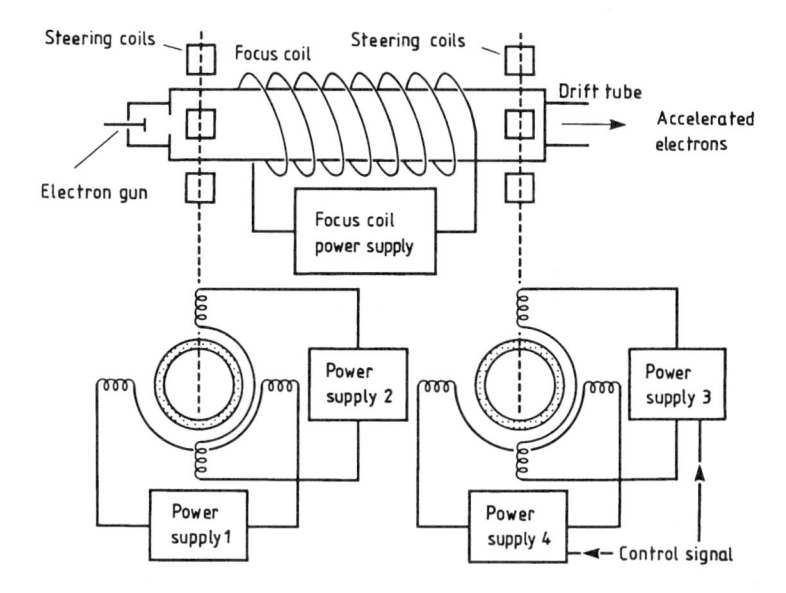

**Figure 5.5.** *The disposition of steering coils around linear accelerator.*

steering fields to optimal values results in some, and in the worse case all, of the electrons being lost from the beam, with a consequent loss in radiation output and a large increase in unwanted x-ray emission from the parts of the accelerator inadvertently struck by stray electrons.

After the electrons have been accelerated to near their maximum energy the second set of steering coils is used to guide the beam accurately onto the x-ray target (or electron window) either directly, in the case of an 'in line' treatment head, or via evacuated drift tubes and bending chambers for the more common arrangements discussed below in section 5.3.3.

Both the direction of the electron beam and its position, as it strikes the x-ray target critically, affect the dose distribution in the x-ray field. To achieve the specified stability of this dose distribution, the power supplies to the steering coils at the high-energy end of the accelerator are preset to provide currents close to the optimal levels but are then continuously controlled by correction signals taken from sensors in the radiation field. This is one aspect of servo control of beam uniformity, which is discussed further in chapter 7.

## 5.3.2. Focusing coils

As the electrons are accelerated through the guide they are subject to forces that will tend to make the beam diverge. There is a small radial component to the microwave field in the waveguide and the individual electrons, being particles of the same charge, will repel each other. As the

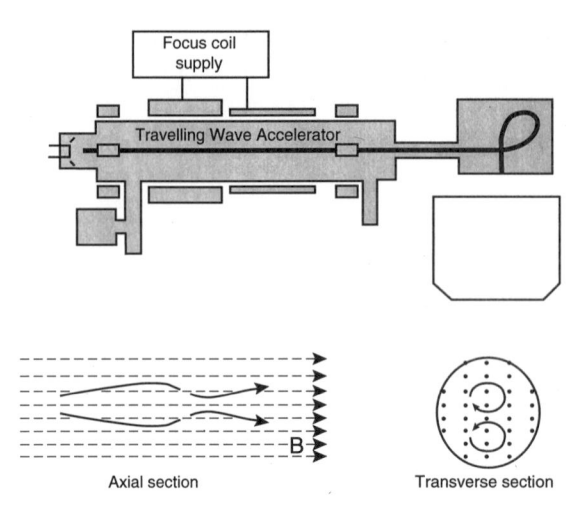

**Figure 5.6.** *An explanation of convergence in a solenoid field.*

electrons gain energy and momentum the defocusing effect of these forces decreases. The radial force from the electric fields remains constant but as the momentum increases beam becomes more rigid and the resultant deflection decreases continuously. Secondly as they reach relativistic velocities the magnetic attractive forces, between the individual electrons forming the beam, counteract the divergent forces. The strongest focusing is therefore required at the low-energy end of the guide. and in some cases it is possible to dispense with focusing in the high-energy part of the guide. The focusing fields are provided by a series of solenoids whose effect on electrons has been explained at the simplest level in section 5.1.2. For clarity figure 5.5 shows just one of these solenoids. At this simple level focusing with a solenoid field would be a misnomer as it would prevent divergence but not cause convergence.

However, in practice, the solenoid does cause convergence at points either inside or beyond the coil. Figure 5.6 shows two electrons which are initially on divergent paths in the solenoid field. The upper electron will follow a clockwise helix but the lower electron, whose vertical component of velocity is opposite, will follow an anti-clockwise helix. Both electrons are in phase and both have the same rotational frequency so after one revolution each helix will be tangential to the axis, as noted in the short section on beam optics, and the electrons will converge. The precise point of convergence depends on the magnetic field strength, the velocity of the electrons and the microwave field strength (which is responsible for the radial electric field), all of which vary along the length of the accelerator.

Computation of electron trajectories is beyond the scope of this book, but some gross approximations can be used to provide a first-order solution.

Assuming that the length of an accelerator, $L$, is 1.5 m, that the focusing magnetic field is uniform and that the beam converges at the exit. The average velocity, $V$, of the electrons in a 4 MeV accelerator is close to $c = 3 \times 10^8$ m s$^{-1}$. In order to produce convergence it is necessary for the reciprocal of the transit time to be equal the frequency of helical rotation ($f = (B \times 2.8 \times 10^{10}) = V/L$). Evalutaion of this approximation gives a field strength $B$ of 7 mT, which is about one-quarter of the average field strength that is actually required, but this is not surprising since we have ignored the other divergent forces that have to be counteracted in practice.

The solenoid coils are coaxial with and outside the accelerating guide and its water jacket, which clearly have to be made of non-magnetic materials. The increased magnetic field at the low-energy end of the waveguide can be provided by having more turns per unit length at this end, or usually by having a number of coils arranged in line along the axis, each providing a field in the same direction: in this case the current in each coil and hence the field in each part of the waveguide can be controlled separately.

As with the initial steering fields, the focusing field performs two related functions: it stops the electron beam from diverging and striking the accelerating guide, and it brings the electron beam section to the required size at a specified point in the drift tube beyond the guide. The precise currents in the coils are determined empirically and need to be very well stabilized and monitored so that the accelerator is automatically shut down if the currents goes outside acceptable limits. A malfunction or maladjustment of these supplies can result in unwanted x-ray production by electrons striking the accelerating guide, or can change the radiation dose distribution inside the useful radiation beam.

The geometrical relationship of the focus coils to the accelerating guide needs to be well defined, so the coils have to be formed on a rigid structure which is firmly fixed with respect to the guide.

In order to generate the necessary high fields solenoids can be constructed either from a few turns in which case very high currents are needed or from many more turns, in which case the currents are lower. In each case the power dissipation in the coils is considerable and forced cooling, from the chilled water system described in chapter 4, is necessary.

### 5.3.3. Beam bending

For accelerators operating at electron energies above 6 MeV the accelerating waveguide cannot be set in line with the central axis of the treatment beam, as to do so would result in an unacceptably high isocentre. The solution is to mount the waveguide parallel or at an angle up to 30° away from being parallel to the gantry axis of rotation, so that the electron beam must be bent to bring it onto the x-ray target or electron beam window. There are many

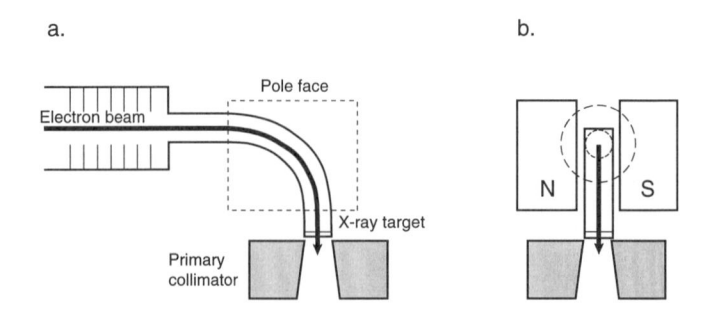

**Figure 5.7.** *The 90° bending system: (a) in the plane of the electron orbit; (b) normal to the plane of the electron orbit.*

ways to achieve this bending but all fall into one of the three categories described below.

*5.3.3.1. 90° bending.* The term '90° bending' should not be taken too literally: it is used to describe bending through any angle close to 90° (usually slightly more), by a simple magnet with the characteristics described here. The main features of the 90° bending system are shown in figure 5.7. Electrons emerge from the accelerator into a drift tube and then enter a flat vacuum box which is between the plane parallel poles of a dipole magnet. For a single-energy machine this may be a permanent magnet, but it is more commonly an electromagnet to allow for the use of different electron energies. The field strength is chosen to bend electrons of the required energy through precisely the angle between the waveguide and the treatment beam axis. As has been explained earlier, the electron beam travels in a circular arc in the field of a magnetic dipole and is then incident on the x-ray target or the thin metal foil window. The radius of curvature of the electron beam is dependent on the electron energy and therefore the 90° bending system acts as an energy spectrometer, the higher-energy electrons being bent somewhat less than those of lower energy. It is therefore necessary to stabilize both the magnetic field and the electron energy very precisely. Even if this is done the narrow energy spectrum of electrons emerging from the accelerator is dispersed by the spectrometer effect. As a result the beam, and therefore the focal spot on the target, are elongated in the longitudinal direction and so the focal spot is ellipical rather than circular. A further second-order effect arises from the slightly different path lengths in the magnet field for electrons with different displacements from the beam axis as they enter the bending magnet. Figure 5.8 shows that an electron that takes a longer path (dotted line) in the field will be bent more than one that takes a shorter path (solid line). The two electrons shown will be focused at the point where they leave the magnet but will be travelling on divergent paths. A similar but complementary effect occurs for angular displacments at the entrance

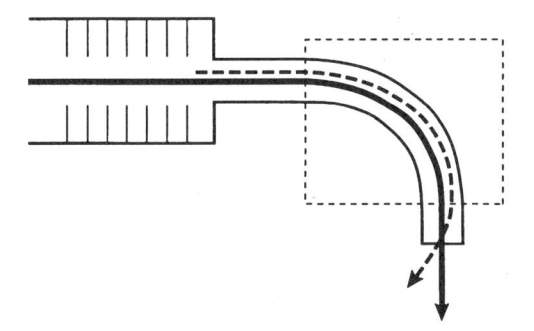

**Figure 5.8.** *The effect of displacement at the entrance to 90° bending system.*

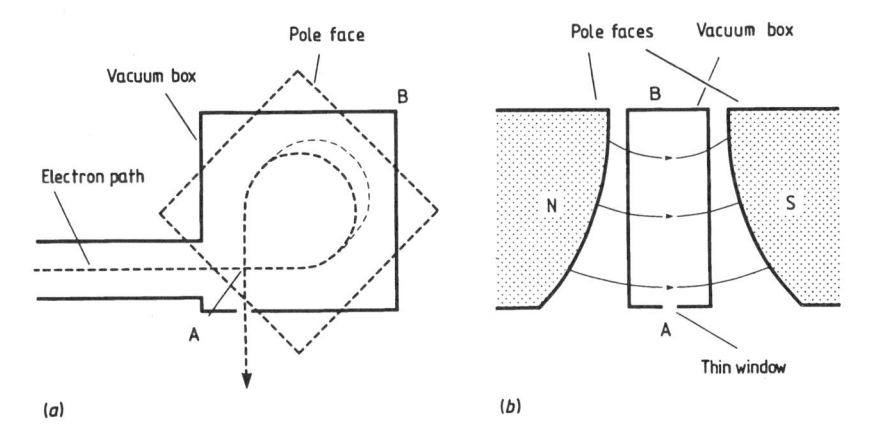

**Figure 5.9.** *The 270° bending system for an electron beam: (a) in the plane of the electron orbit; (b) normal to the plane of the electron orbit.*

to the bending magnet. It is therefore important to ensure that the beam is steered precisely into the bending chamber, entering it at the correct position and direction.

*5.3.3.2. 270° bending.* The purpose of a 270° bending system is to accomplish achromatic bending, ideally so that electrons will strike the x-ray target or pass through the window at the same point and in the same direction, independent of their energy.

Figure 5.9 shows one example of a 270° system first proposed by Enge (1963). Here the electron beam follows paths as shown in figure 5.9(a). The field is provided by an electromagnet whose pole faces normal to the plane of the beam, shown in figure 5.9(b), are shaped to give an approximately hyperbolic pole gap. In the plane of the electron beam the magnetic field strength increases radially about point A. As an electron penetrates into the

68     *The electron beam (its production and transport)*

magnetic field it will follow a path whose curvature depends on its energy and the local field strength. As more energetic electrons will penetrate further into the magnetic field than those of lower energy, they follow a larger orbit. However, having done so they will be subject to a stronger field. The achromatic property of this arrangement can be explained by making the following observations. The magnetic field and hence the orbits are symmetric about the line AB. During the first half of the orbit all electrons are deflected through 135° but are dispersed along AB depending on their energy. During the second half the electrons converge to emerge from the deflecting field at the same point and travelling in the same direction.

The variations in the magnetic field strength have been chosen such that all such orbits have the same shape and so that electrons of any energy cross the line AB at 90° this being the necessary condition for symmetry of the orbits (which cannot have discontinuities in direction).

This discussion so far applies to electrons travelling in the mid-plane of the electromagnet gap. It can be seen from figure 5.9(b) that the magnetic lines of force are necessarily curved, and this curvature provides a force towards the mid-plane on any electron diverging from this plane. In other words the magnetic field has a focusing effect in the plane of figure 5.9(b). (A magnet of the type just described is sometimes called a 'pretzel' magnet, for the rather unlikely reason that the path of the charged particles through its field is the same shape as the German biscuit of the same name.) A useful corollary of the achromatic bending system is that, as long as the gradient is maintained, the absolute magnitude of the field is not critical, so if the current in the bending magnets falls to slightly below its design level then the only consequence will be that the electrons follow a slightly larger orbit. However it should be noted that the acceptance range for electron energy is limited by the size of the bending chamber and pole pieces, which determines the highest energy, and by the accuracy of the field gradient, which determines the lowest energy that can be successfully bent.

As has been stated at the beginning of this section the achromatic system described above is just one of a number of alternative arrangements for producing 270° achromatic magnetic bending systems, some of them using multiple magnets others with differently shaped poles.

*5.3.3.3.   112.5° bending.*   Neither of the two bending systems described above is ideal. 90° bending is dispersive and while the 270° bending solves this problem it is at the expense of an increase in the overall height of the treatment machine.

The 112.5° bending system shown in figure 5.10 addresses both these problems. That is, it is achromatic and avoids the increased height that arises from the beam trajectory in a 270° arrangement such as that in figure 5.9.

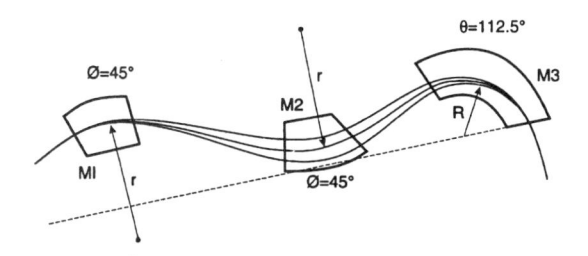

**Figure 5.10.** *The 112.5° bending system for an electron beam.*

There are three sectors along the line of the evacuated flight tube into which electrons are launched from the accelerator. The first sector deflects electrons of the required energy through 45° and acts as a spectrometer: higher-energy electrons are deflected by less than 45° while lower-energy electrons are deflected more than 45°. The electrons, which are now diverging as a result of their energy spread, enter a second sector where they are again deflected by 45°, but in the opposite direction. During this part of their trajectory they begin to converge. The final sector bends the beam through 112.5° and completes the energy convergence so that all the electrons emerge at the same point and travelling in the same direction. It should be noted that although the overall effect of this system is to bend electrons through 112.5° the sum of the deflection angles (without regard to the sign) is actually 202.5° which is not very much less than the alternative achromatic arrangement described above.

Although the main purpose of the slalom system is to produce achromatic bending without requiring the trajectory of electrons to extend far above the target there are other more subtle features described by Botman *et al* (1985): these include the focusing of the beam in both its cross sectional directions so that the focal spot on the x-ray target is typically 2 mm.

The sectors are electro-magnets whose fields can be adjusted to transport electrons of a particular energy; the acceptance window is about ±10% of the selected energy, which is sufficient tolerance for the energy spectrum produced by the accelerator.

## 5.4. THE COMPLETE ELECTRON ACCELERATING SYSTEM

In this and the previous chapters all the sub-systems that comprise the complete electron accelerating system have been described. They are brought together into a composite diagram in figure 5.11 in order to show their interrelationship. In this example, which is not intended to represent a particular make or model, there is a magnetron driven travelling waveguide with recirculation of microwave power through a four-port hybrid coupler.

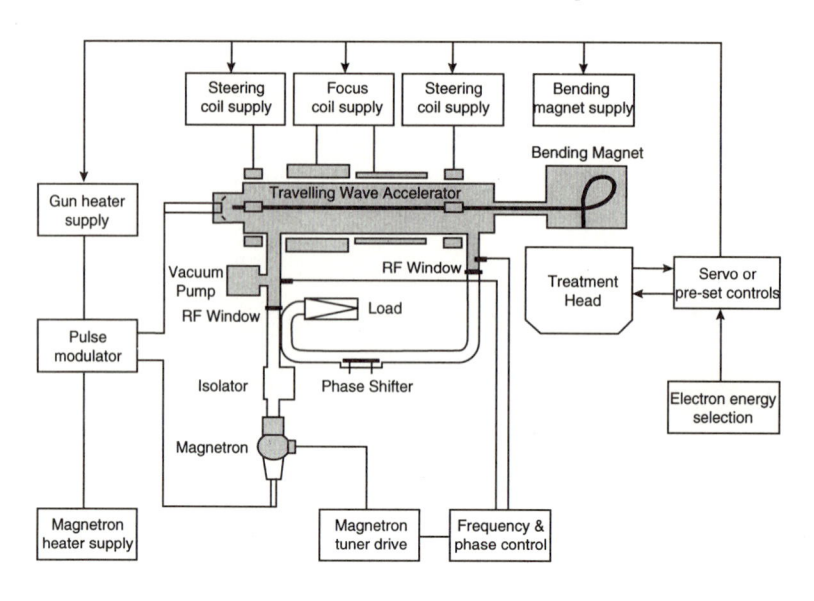

**Figure 5.11.** *A block diagram of the complete acceleration and transport system.*

The magnetron tuner is controlled to give the optimum microwave frequency and phase conditions in the waveguide.

With the exception of the pulse modulator, which has been described in some detail in chapter 3, the need for electrical power supplies for each of the sub-systems has been assumed. The first set of steering coils, the focus coils and those energizing the 270° bending magnet all require well stabilized DC power supplies. These are shown here together with schematic representation of some of the more important feedback control signals particularly to the diode gun heater supply and the second set of steering coils where in both cases the control signals are derived from monitoring points in the treatment head and beam monitoring systems that will be described in the subsequent chapters. The sections of waveguide that are either evacuated or pressurized are shown but the water cooling system which is ubiquitous has been omitted for clarity.

# CHAPTER 6

# THE TREATMENT HEAD

The treatment head is that part of the machine which receives the accelerated electrons from the system described in the previous chapter and uses them to generate either an x-ray or an electron beam for treating patients.

There are two configurations of the treatment head relative to the accelerating waveguide: these are shown in figure 6.1. The treatment head can be either in line with the waveguide in which case the target is permanently fixed to the end of the flight tube, or at right angles in which case a beam bending magnet (chapter 5.3.3) is required between the end of the flight tube and the target. A more detailed diagram of the treatment head for a simple machine is given in figure 6.2. The inner cone provided by the primary collimator projects onto the edges of the target and defines the largest available field size. Its external dimensions are normally chosen to attenuate any x-rays not passing through the inner cone to less than 0.2 per cent. The primary collimator is either a lead filled steel casting or a heavy-metal, tungsten–copper, alloy. Where space is very limited depleted uranium can be used as the shielding material.

The beam flattening filter, designed to give a uniform dose distribution across the x-ray field is shown mounted on the end of the primary collimator. The beam monitor is also fixed with respect to the primary collimator. The beam monitor, a set of ionization chambers which monitor dose rate, dose and dose distribution in the field, is mentioned here for completeness but is described in more detail in chapter 7.

In this type of machine the target, primary collimator, field flattening filter and beam monitor can all be rigidly mounted in relation to the main frame of the accelerator structure to which the support plate of figure 6.2 is fixed. The remaining components of the x-ray head, the movable beam defining collimators, the mounting for wedge filters and the light field system all need to be able to rotate about the central axis of the x-ray beam. These are contained within the rotating frame, which is a substantial structure mounted on a large bearing. It carries, at its patient end, an accessory ring which provides a rigid mounting for any necessary mechanical or optical beam direction devices.

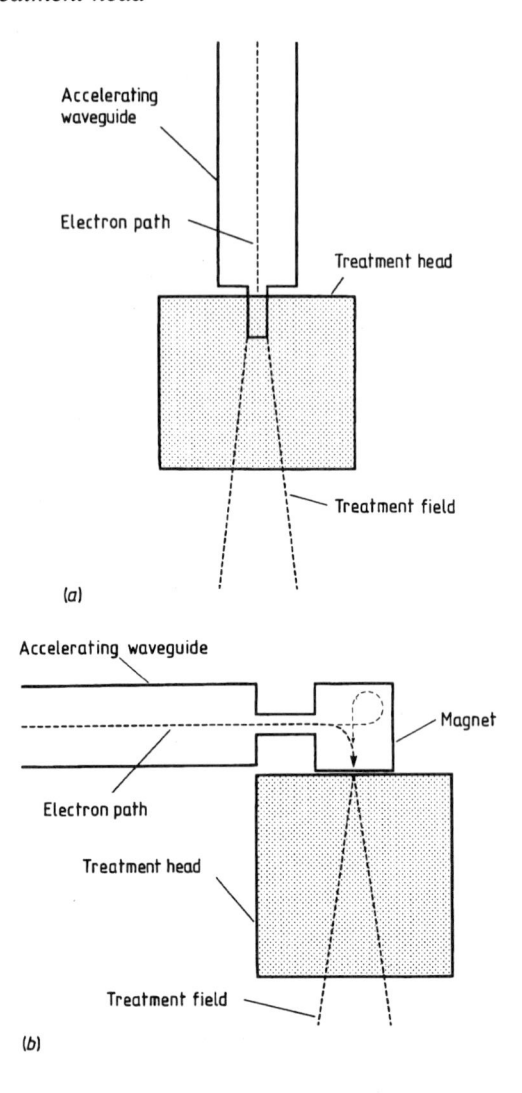

**Figure 6.1.** *The arrangement of the treatment head in relation to the accelerating waveguide: (a) treatment beam in line with the accelerating waveguide; (b) treatment beam perpendicular to the accelerating waveguide.*

The movable beam defining collimators, again made of a lead, tungsten or a similarly dense alloy, are mounted in pairs such that their inner surfaces project onto the edge of the x-ray target, thus minimizing the width of the penumbra on the radiation field. This system will define a continuously variable range of rectangular fields, up to a maximum size defined by the inner cone of the primary collimator. On current machines, field sizes up to $40 \times 40$ cm$^2$ at 1 m from the target (in the isocentric plane) are available.

*The treatment head* 73

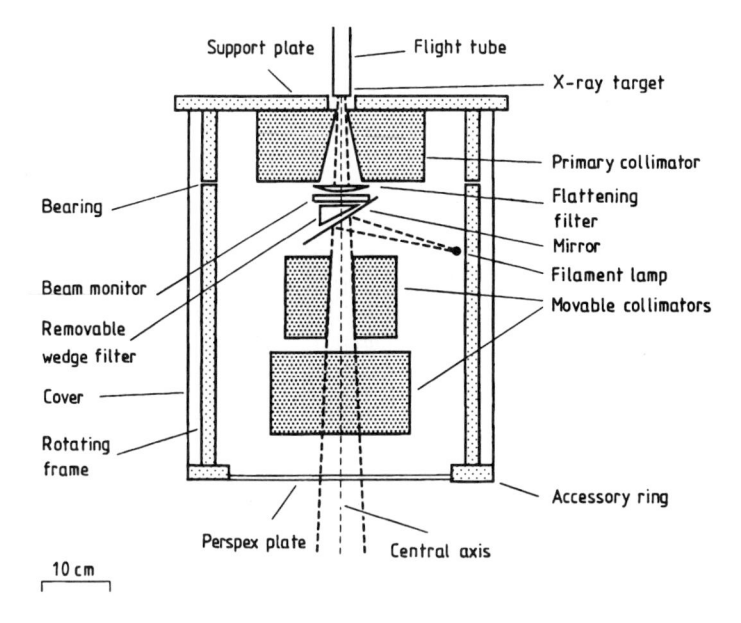

**Figure 6.2.** *The treatment head for the straight electron beam system.*

It may be noted that the relationship between collimator setting, i.e. the spacing of the collimators, and field size is not unique, being a function of source to skin distance (SSD), and also because, in a situation where the edges of the field are not perfectly sharp, the definition of field size is a matter of convention. This will be discussed later in chapter 8.

The optical beam system to illuminate the field defined by the collimators is illustrated. To simulate the x-ray field accurately the distance of the light source from the mid-point of the mirror has to be equal to the distance from the x-ray source to the mirror. In order to keep down the overall diameter of the system about the central axis, it may be necessary to fold the optical path by the use of an additional mirror or mirrors, or to physically shorten it by the use of a convex lens which gives a virtual source at the correct distance.

Removable wedge filters may be mounted at the position shown, which allows the use of the optical beam, or at the patient end of the system, with respect to the accessory ring.

Linear accelerators are normally used either with the patient's surface at 1 m from the x-ray source (fixed-SSD treatments) or with the centre of the volume to be treated at this distance (isocentric treatments).

The geometric penumbra is related to the size of the x-ray source (the focal spot), which may be as small as 2 mm in diameter. In order to minimize the penumbra of the x-ray beam, the beam defining collimators should be as near the surface of the patient as possible. On the other

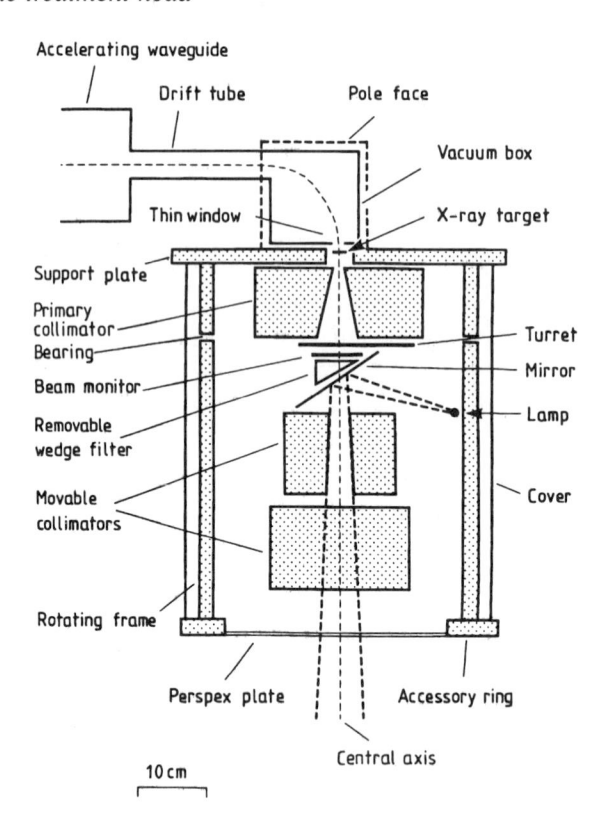

**Figure 6.3.** *The treatment head for the bent electron beam system.*

hand to have reasonable access to the patient there needs to be adequate clearance between the accessory ring and the patient's surface. Further, by placing the beam defining jaws as near the target as possible, the amount of movement needed to give the required range of field sizes is minimized as is the weight of each of the collimating blocks. It is also desirable to keep the beam defining system, and indeed any material in the x-ray head, as far as possible from the patient's skin, to minimize the dose component due to scattered electrons. The compromise between these conflicting requirements is to place the accessory ring at 50–60 cm from the target.

The thin plastic window which, if fitted, closes off the end of the beam defining system can serve two main purposes; it can indicate the central axis by projection of inscribed marks by the the optical beam system, and it prevents dust from entering and items from being dropped into the treatment head. The accessory ring is in a plane normal to the central axis of the beam, its outer surface being concentric with the central axis, and serves as a mounting for beam direction devices.

In a more complex machine more than one x-ray energy may be available and electron beams of several pre-set energies may be extracted. In this case a flattening filter for each x-ray energy and scattering foils for each electron energy must be provided. These are mounted within the treatment head and are driven into the correct position when required. These devices are shown in figure 6.3 as being mounted on a rotating turret. On the current generation of machines the mechanical movements of the x-ray head, the rotation of the head about its main bearing, the adjustment of the beam defining collimators, the movement of the x-ray target and the rotation of the turret carrying the flattening filter are all motor driven.

## 6.1. THE X-RAY TARGET

Transmission targets are employed because in this energy range the photons produced are directed mainly in the same direction as the incoming electrons. Since the efficiency of x-ray production increases very rapidly with increasing electron energy for machines operating in the megavoltage range, target heating is not a serious problem and can be controlled by cooling water flowing through a copper block into which the target is fitted. For a given electron energy the photon spectrum generated depends on the atomic number and the thickness of the target, thickness being expressed in terms of electron range in the material of the target. The mean photon energy will be greater for a thin target, where the average energy of the electrons interacting by the bremsstrahlung process will be greater than those in a thick target, but there will also be less x-ray output. For a thin target there will be an unwanted flux of electrons on the patient side of the target, and these can be absorbed in a low-atomic-number material on the back of the target, such as carbon, with minimal x-ray production. Podgorsak *et al* (1975) have looked experimentally at the radiation quality and output produced by different targets and concluded that for electron energies up to 10 MeV a thick tungsten target gave the best compromise between good x-ray output and beam penetration, while for higher electron energies a thick aluminium target should be used.

## 6.2. THE BEAM FLATTENING FILTER

The unfiltered x-ray beam from a megavoltage generator gives a sharply peaked dose distribution along lines normal to the central axis of the beam. Figure 6.4 gives an example of the x-ray dose distribution from an 8 MV linear accelerator (curve A), measured at 1 m from the target in the isocentric plane. Under ideal conditions, where the electron intensity distribution on the target is perfectly uniform and when all the electrons arrive along lines normal to the target, this curve would be symmetrical. The edges of this curve are defined by the collimators in the x-ray head. This dose distribution

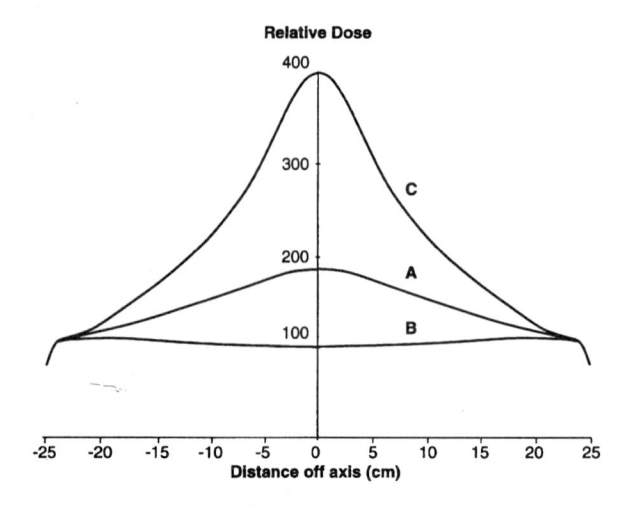

**Figure 6.4.** *Measured dose distributions along a line perpendicular to the central axis of x-ray beams, 1 m from the target. A, 8 MV x-ray beam without flattening filter. B, 8 MV x-ray beam with flattening filter. C, 20 MV beam without flattening filter.*

becomes even more peaked as the electron energy is increased, as is shown by curve C for 20 MV x-rays.

The main function of the filter in a megavoltage x-ray generator is to 'flatten' these curves. It is a cone shaped device, which differentially absorbs the radiation towards the beam centre. The effect of the flattening filter for 8 MV x-rays is shown in curve B.

This flattening filter substantially reduces the dose rate at the beam centre, and it can be seen that the relationship between the unflattened and flattened dose rates on the central axis of the beam will depend on the maximum field size for which the flattening filter is to be used. At high energies, as can be seen from the 20 MV curve, the flattening filter has to reduce the dose rate at the field centre very substantially for large fields. The dose rates available from linear accelerators are sufficiently high to allow this loss of dose rate, and still have a useful output in the flattened field of the order of hundreds of centigrays per minute at 1 m from the target.

The beam flattening filter also acts as a radiation filter in the more traditional sense, i.e. by differential photon absorption it can change the radiation spectrum. As can be seen from figure 6.5, this process in a high-atomic-number material such as lead may actually differentially filter out the high-energy photons in the spectrum, for a machine working for example at an electron energy of 30 MeV. In order to avoid softening of the beam in this way it is desirable to use a medium-atomic-number material, and it can be seen that aluminium is ideal for this purpose. This is in agreement with experimental studies by Podgorsak *et al* (1975), who showed that maximum

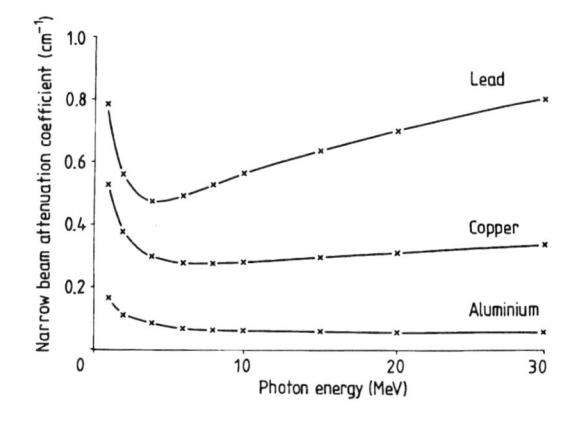

**Figure 6.5.** *Narrow-beam linear attenuation coefficients for x-rays as a function of photon energy. Data are from the book by Hubbell and Berger (1968).*

depth doses in water are given by the use of an aluminium flattening filter.

A first approximation to the design of a beam flattening filter can be made by determining the shape of the unflattened curve (figure 6.4), and then using attenuation coefficients to determine the required thickness of material to differentially filter the beam at each point on the curve. This is only an approximation because photon scattering in the flattening filter will contribute to the measured dose and because the energy of the x-rays generated at an angle to the incident electron direction is somewhat lower than that generated in the forward direction. As a result the final design has to be determined by successive approximations. So far in this discussion the dose distributions 'in air' have been considered. In practice, it is the dose distribution in a phantom which is of interest, so the situation is further complicated by scattering processes in the phantom. It is then only possible to have a flat beam at one depth. For instance, if it is decided that the dose distribution is to be flat at 10 cm deep, this may require some 'over-flattening' at depths nearer the surface and 'under-flattening' at greater depths.

At the higher energies, the required beam flattening filter will be several centimetres thick at the centre and finding adequate space for it in the x-ray head (figure 6.3) may become a problem. It is then necessary to compromise between the considerations which led to the suggestion that aluminium was the optimum material, and to construct a filter which will fit into the space available. For this reason copper or stainless steel, which have reasonable properties in terms of differential photon energy filtration, are often used as compromise materials.

The position of the flattening filter is critical with respect to the central axis of the beam; any movement of either relative to the other will make an otherwise flat beam unflat and asymmetrical. In other words, the mechanical position of the flattening filter has to be well defined, and the position and direction of the radiation beam have to be well stabilized. These latter factors are controlled by signals from the beam monitoring system.

Clearly the shape and size of the flattening filter are very strongly dependent on the x-ray energy. Linear accelerators that are able to operate at two alternative energies must therefore have two corresponding flattening filters, which must be carefully interlocked with the operating conditions. A second consequence of the energy dependence is that small changes in x-ray energy cause the beam to be either over- or under-flattened. The necessary monitoring of beam uniformity will be discussed in chapter 7.

An alternative to the use of a beam flattening filter is to scan the electron beam to strike the target from different directions. This is not a common technique and is mentioned here only for completeness. The rationale is that at very high energies the polar diagram of x-ray production relative to the direction of the incident electron beam is so narrow that the flattening filter necessary to produce a sufficiently wide beam would be impractically thick. By arranging to modulate the electron beam direction at the point where it strikes the target the direction of the narrow lobe of radiation produced will also be modulated. Summation of a series of lobes with different directions produces an effective polar diagram which is uniform in all directions covered by the useful primary beam.

## 6.3. PRIMARY COLLIMATOR AND BEAM DEFINING COLLIMATORS

The thickness of shielding material, usually lead alloyed with other metals to improve its mechanical and machining properties, required in these elements of the beam defining system does not vary greatly over the photon energy range used in linear accelerators. This follows from the slow variation of the attenuation coefficients as a function of photon energy (figure 6.5). The mean photon energies for linear accelerators operating in the 4–30 MV ranges are about 1.5–10 MeV and in this range the narrow-beam attenuation coefficients may be passing through the minimum value associated with the onset of pair production. It should be noted that these values are only quoted as examples, the attenuation in the beam defining system being nearer to 'broad-beam' rather than narrow-beam conditions. The actual thicknesses of material required have to be determined empirically and these are roughly indicated by the scales on figures 6.2 and 6.3.

The mechanical arrangements for moving the beam defining collimators, while maintaining the condition that their inner edge projects on to the target edge, are usually quite complex and will not be described in detail here. The requirement is that the collimator block should move in an arc centred on the target. Failure to meet this requirement results in not only a sub-optimal penumbra but also one which varies with collimator setting. The amount of backlash that can be tolerated in the collimator mechanism is very small. For example, a backlash or flop of 1 mm in the collimators nearest the target may move the beam edge and also the central beam axis of symmetry by up to 0.5 cm in the isocentric plane and this would not be tolerable. The collimator positions also need to be accurately indicated, either mechanically on a scale, or via an electrical transducer, and again the maximum backlash between collimator position and indicator has to be less than a fraction of a millimetre. It should be noted here that although each collimator is inside the treatment head, maybe 35–40 cm from the target, the 'collimator position', which will be referred to frequently, is the position of the 'projection of the collimator' onto the isocentric plane which is 100 cm from the target.

### 6.3.1. Symmetric collimators

In most older machines the beam defining collimators were designed to move in pairs symmetrically about the axis of rotation of the x-ray head, one of the axes defining the isocentric mounting of the machine. The geometry of the isocentric mountings is described in chapter 9. As has been stated earlier the penumbra in the isocentric plane is related to some extent to the size of the focal spot and the relative position of the collimators. As one pair of collimators will be closer to the target than the second, the geometric component of the penumbra for these inner collimators will be greater than for the outer set. To offset this difference some systems make use of penumbra trimmers, which are relatively thin additional collimators mounted as outriggers onto the inner collimators so that they move in a plane just beyond the outer collimators. In this way it is possible to ensure that the penumbra for each set of collimators is, for practical purposes, the same.

### 6.3.2. Independent collimators

Further flexibility can be achieved by controlling the position of each of the four collimator blocks independently. The same requirements for focusing each block onto the target still apply and of course the number of controls and position transducers doubles. Each of the collimators can move from the central axis to the furthermost open position, projected to 20 cm off axis in the isocentric plane. In some designs all the collimators can move across the axis by up to 10–15 cm but others have mechanical constraints such that only one pair can cross the axis by a significant amount. The use of independent collimator designs, particularly where they can cross the central axis, are usually incompatible with the use of penumbra trimmers as the mechanical connection of the trimmer to the main inner collimator is not possible.

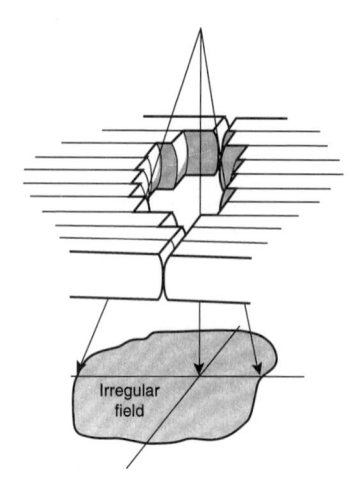

**Figure 6.6.** *A simplified diagram of an MLC.*

The most common application of independent collimators is to produce one-half or three-quarters blocked fields where one or two of the beam edges are coincident with the beam axis. As there is no beam divergence along this axis it is possible to match beams from different directions without concerns about overlaps as the beams diverge (e.g. it is frequently necessary to match the upper edge of parallel tangential beams irradiating the breast with the bottom edge of an anterior field irradiating the supraclavicular region). Conventions for describing the size and position of asymmetric rectangular beams will be discussed later but is should be noted that the addition of two further variables carries with it the potential for compounding of small errors in field size measurement and control, thus the accuracy requirements are somewhat higher than for symmetric systems.

### 6.3.3. Multileaf collimators

The production of radiation beams with shapes other than rectangles is often necessary so that the volume of tissue irradiated can be minimized and so that sensitive normal tissues close to the treatment field can protected. Although beam shaping with shadow blocks, which will be described in chapter 8, is possible on all machines, the automatic shaping of beams has become a requirement on many machines. Multileaf collimators (MLCs) have been developed to meet this requirement and the design and performance characteristics of one such device is descibed by Jordan and Williams (1994).

Figure 6.6 is a simplified diagram of an MLC consisting of two opposing banks of attenuating leaves, each of which can be positioned independently. The leaves must be sufficiently thick to provide the necessary attenuation and sufficiently narrow to provide the necessary spatial resolution in the direction

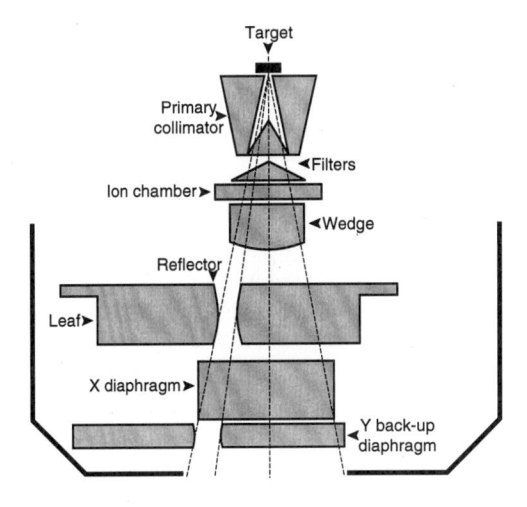

# Primary collimator and beam defining collimators

**Figure 6.7.** *A schematic diagram of an MLC integrated into the treatment head.*

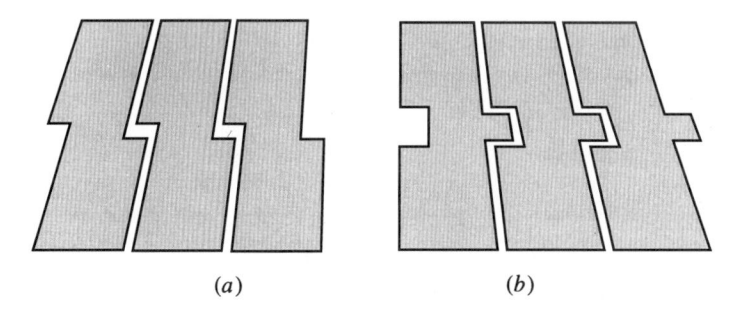

**Figure 6.8.** *Two methods of reducing leakage between adjacent leaves. (a) Steps. (b) Tongues and grooves.*

normal to the leaf movement: this is usually 1 cm in the isocentric plane. The spatial resolution in the direction of movement of each leaf is determined by the precision of the control system and this is usually better than 1 mm. There have been many different designs of MLC falling into three main categories: those in which the leaves provide the only collimation (apart from the primary cone), those in which the leaves are mounted externally to a standard treatment head and those in which the leaves and back-up collimators replace the standard treatment head. The following description centres on the replacement type of system but most of the points considered are applicable to the other configurations.

In this example, shown in figure 6.7, the two banks of tungsten leaves are mounted inside the treatment head immediately below the flattening filter and housing for the motorized wedge, which will be described in the

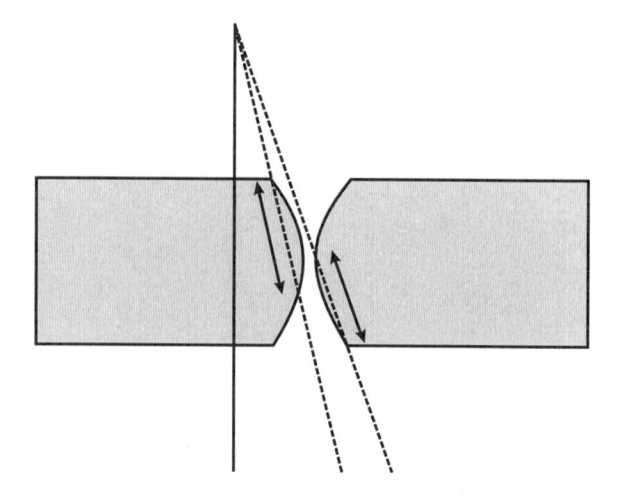

**Figure 6.9.** *Collimator leaves with curved faces. The lines show partial transmission resulting in degradation of the penumbra.*

following section. In this position the leaves are at approximately one-third of the distance from the x-ray source to the isocentre and are therefore approximately 3 mm wide in order to project to a leaf width of 1 cm in the isocentric plane; the leaves are tapered so that each of their edges can be focused toward the target. The leaf thickness, of about 7 cm, is sufficient to reduce the intensity of the primary beam to 1%. Additional attenuation is provided by the back-up collimators, the requirements of which will become apparent after further consideration of the leaves. In this design the leaves are mounted on rollers which allow them to be moved across the beam by lead screw mechanisms driven by small DC motors. There has to be a mechanical clearance between the leaves to permit easy movement and this results in leakage between adjacent leaves, which is minimized by the use of stepped and overlapped leaf sections as shown in figure 6.8(a) and by keeping the gap as small as reasonably possible. An alternative method of overlap analagous to a tongue and groove joint is shown on figure 6.8(b). As long as the overlap is maintained the maximum leakage between leaves would be approximately 10% but in practice lower leakage is achieved because the gap, being very small, does not subtend the full focal spot from points in the isocentric plane. The average leakage through banks of leaves designed in this way is about 2% with local maxima of up to 5%. Such levels of leakage over the full area exposed by the primary collimator are not acceptable and the back-up collimator moving in the same direction as the leaves provides attenuation by a further factor of ten so that areas shielded by both leaves and back-up collimators receive a dose of less than 0.5% of the dose in the beam. A more difficult leakage problem arises between the opposing banks of leaves. When a pair of opposing leaves is required to be closed the gap, for mechanical clearance, is exacerbated by the curved profile, shown in figure 6.9, the need for which will be explained below. The projection of this gap onto the isocentric plane can be reduced slightly by setting the closed position away from the central axis but the leakage will approach 100%: as a result the back-up collimator moving normal to the leaves must provide attenuation to about 1% of the open beam intensity.

The curved leaf profile provides a method of minimizing the penumbra without the complex mechanical arrangements necessary to keep the flat edge of a conventional collimator aligned with the x-ray source. Although the curvature allows some penetration of the beam through the edge the depth of penetration of the ray corresponding to 20% transmission (and hence the degradation of geometric penumbra) is independent of the position of the leaf as shown by figure 6.9.

Clearly the positional control of up to eighty-four moving elements of a multileaf collimator (up to eighty leaves and four back-up collimators) is somewhat more complicated than for simple symmetric or independent systems. It is beyond the scope of this section to describe the control system in great detail but it should be noted that the overall control system must include multiple control channels each with the facility to measure the position of each element and a motor to change its position. The measurement of the position of each leaf can be by analogue or digital transducers (potentiometers or shaft encoders) or by more sophisticated means such as the optical and computer vision based system which is described in more detail as an example of computer control technology in chapter 10.

## 6.4. WEDGE FILTERS

There are several reasons why it might be necessary to produce a gradient across the x-ray fields. These include the matching of gradients from combined fields to achieve dose uniformity throughout a target volume and compensation for oblique incidence to provide a uniform dose in a defined plane within a patient. Wedge filters, as their name implies, were originally metal wedges designed to produce the desired dose distributions, as illustrated in figure 6.10(a). Beams modified by wedge filters are often described as wedged beams. The wedge angle, while not a full description of the overall effect of wedge on the dose distribution, is used to quantify the effect of each wedge filter and is loosely defined as the angle between the central axis and the normal to a defined isodose. That is, it measures how much the isodose curves have been tilted by the wedge as shown in figure 6.10(c).

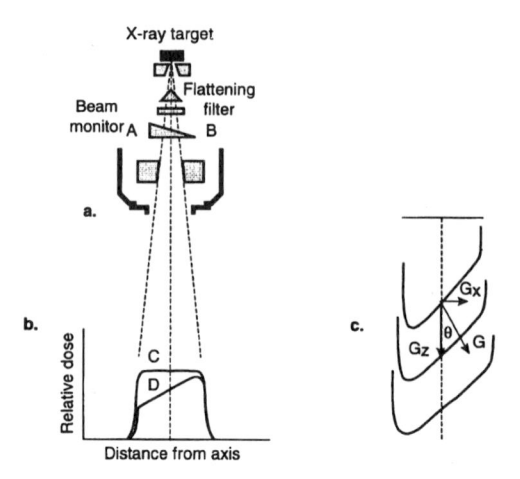

**Figure 6.10.** (a) The position of a wedge filter in treatment head. AB is the critical direction for positioning the filter. (b) The effect of the wedge filter on dose distribution across the x-ray field: C, normal dose distribution; D, dose distribution with the wedge filter in position. (c) Isodose curves for a wedged beam showing the dose gradient G and its components $G_z$, along, and $G_x$, normal to, the central axis.

There are several variants of the wedge filter which will be discussed together with an alternative method of producing a wedge beam which is widely used.

### 6.4.1. Removable wedges

Removable wedges are individually designed metal wedges, each of which is shaped to give a particular required dose distribution at a specified depth in a phantom. In order to produce a constant gradient across the beam the filter does not have the simple triangular section as shown. The complex shape required is approximately logarithmic but the precise shape is arrived at by a combination of calculation, taking into account the known attenuation coefficients of the material used, and experimental iteration (trial and error), necessary as the calculation does not adequately account for scattered radiation and other second-order effects. The previous discussion about choice of material for the beam flattening filter is also applicable to wedge filters.

Removable wedges may be mounted internally or externally to the treatment head. If they are inside the treatment head, ideally above the main mirror in the optical beam, they are small and relatively light in weight but must be positioned with great precision as small displacements will be magnified at the isocentre. If they are mounted outside they will be heavier and will inevitably obscure the optical beam but they will be clearly visible to the operator and will be more tolerant of small positional errors. It is clear from figure 6.10 that the presence of a wedge filter will change the relationship between the signal provided by the monitor and the dose delivered to a patient (for example, the dose at a point on the central axis of the field) and that this relationship is critically dependent on the position of the filter with respect to the central axis of the beam. It is for this reason that the mechanism for locating the wedge filter needs to be very precise, in the direction of the dose gradient.

Normally, up to six different wedge filters are required to give the necessary range of dose distributions, and it is clear that the presence or absence of a wedge filter, or the use of the wrong one can make a large difference (up to about a factor of four) to the dose delivered to a patient for a given dose monitor reading. To avoid gross errors of this type it is necessary to have an interlock system which will identify the wedge filter in, or on, the treatment head and inhibit the production of radiation if the wedge selection is incorrect.

### 6.4.2. Universal or motorized wedges

An alternative to the use of individual removable wedges each of which is designed for a specific wedge angle, is to use a single wedge which gives the steepest distribution ever needed, and to place it in the beam for a suitable fraction of the exposure. The resultant dose distribution is then the appropriately weighted average of that obtained with the wedge and a uniform beam. The fraction of exposure necessary to produce a wedged beam of wedge angle $\theta'$ from a motorized wedge of angle $\theta$ can be calculated approximately by consideration of the gradients in the dose distribution.

The gradient of the dose distribution at any point is normal to the isodose curve passing through that point.

For a wedged beam the gradient has two components $G_z$ and $G_x$ along the central axis ($z$) and normal to the central axis ($x$).

The wedge angle $\theta$ is given by

$$\frac{G_x}{G_z} = \tan \theta. \qquad (6.1)$$

For two fractions, where $W_1$ is the beam weight for the unwedged fraction and $W_2$ is the beam weight for the wedged fraction, the resultant gradients are

$$G'_x = \left(\frac{W_2 G_x}{W_1 + W_2}\right) \qquad G'_z = G_z. \qquad (6.2)$$

The wedge angle resulting from the weighted beams is given by

$$\tan \theta' = \left(\frac{W_2}{W_1 + W_2}\right) \tan \theta \qquad (6.3)$$

so
$$\theta' = \tan^{-1}\left(\frac{W_2}{W_1 + W_2}\tan\theta\right). \tag{6.4}$$

This assumes that $G_z$ is not affected by the wedge (i.e. no beam hardening) and takes no account of the effects of the wedge on scattered radiation generated in the treatment head or in the phantom. It is important to note that the beam weights relate to the proportion of the dose delivered with and without the wedge in place. If, as is usually the case, the beam is controlled by a beam monitoring system in which the ionization chamber samples the beam before it passes through the wedge, then the proportion of 'monitor units' delivered with the wedge in will be considerably more than the proportion of dose required.

Although universal wedges have been very widely used, either as manually mounted devices or remotely activated and motor driven into position for the prescribed proportion of the exposure, it is necessary to be aware of their limitations compared with the individual removable alternatives.

Dose distributions from wedged fields produced from a single universal wedge will all have the same general shape but those from discrete wedges can be different (for example some wedges might produce slightly convex isodose curves whereas others might be concave). In general small-angle wedges are used for large fields and large-angle wedges are used for small fields. As a universal wedge has to produce both these extremes it must have a large angle and cover large fields. The consequence is that the universal wedge must be significantly thicker (on the central axis) than any of the individual wedges it replaces and will therefore have a higher wedge attenuation factor. If a wedge is of uniform section the oblique section is somewhat thicker than the normal section through the central axis. This is not usually significant for small wedges but becomes so for larger motorized wedges, which have to be profiled in both directions in order to maintain a constant wedging effect.

It is now necessary to consider briefly the use of wedge filters in asymmetric fields in order to draw attention to two points applicable to either removable or motorized wedges.

First, when a wedge is applied to a uniform beam it is highly desirable that the wedge angle should remain approximately constant independent of the degree of asymmetry. This implies that the slope of the wedge (the block of metal) should be constant and that the wedge will have a triangular cross section. Because of the exponential nature of attenuation this gives rise to concave isodoses, which become marked for large beam sizes.

Second, when a wedge has been designed to produce a wedge angle of, say, 50° for a symmetric beam it cannot be assumed that this will be maintained precisely for asymmetric beams. The differences that arise depend on the uniformity of the unwedged beam. By definition the dose gradient normal to the central axis is zero for a flattened beam. Therefore

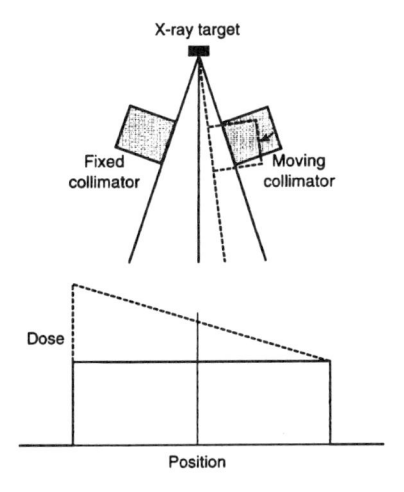

**Figure 6.11.** *Use of the x-ray collimator to produce a dynamic wedge.*

the only gradient in this direction is that imposed by the wedge. However, at the centre of an asymmetric beam, maybe 5–10 cm away from the axis, it is likely that there will be a small gradient depending on the extent to which the beam has been over flattened to compensate for variations in scatter as mentioned in 6.2. These small gradients will add to the wedge angle on one side of the axis and reduce it on the other. The effect can amount to several degrees in the steepest wedge angle and adds yet another variable to be taken into account when beam data is measured and used in treatment planning.

## 6.4.3. Dynamic wedges

The dynamic wedge is the term that has been used to describe the production of a wedged beam by dynamic motion of the collimators during irradiation as illustrated in figure 6.11. The wedged profile can be considered in two parts. First a uniform profile is delivered with the collimators in their starting position determined by the required field size. Then during the second part one of the collimators is driven across the beam at a rate determined by the gradient required. This sort of wedge generation requires precise control of collimators linked to the dosimetry system so that their position is synchronized to the proportion of the exposure given: it is therefore most easily realized on computer controlled linear accelerators. Dynamic wedge production can be extremely flexible as the wedge profile can be programmed to produce any required shape. Furthermore, as the wedge effect is not produced by differential attenuation through a filter the beam is not subject to changes in energy as is the case for physical wedge filters. However the calculation of the collimator trajectory required to produce a

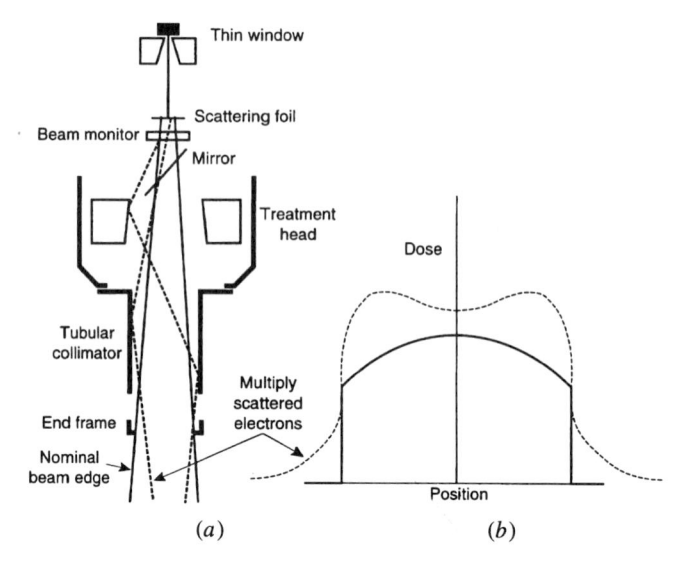

**Figure 6.12.** *(a) An electron beam with a single scattering foil and tubular collimator. (b) The distribution of electrons scattered from the foil and collimator.*

particular wedge profile is not trivial because account has to be taken of scattered radiation as well as radiation in the primary beam.

Dynamic control of both collimators, and the extension of this method to MLCs, will open up many possibilities in beam intensity modulation including the generation of dynamic compensators, the two-dimensional analogue of the dynamic wedge.

## 6.5. ELECTRON BEAM PRODUCTION

Multipurpose linear accelerators are used to provide beams of electrons as well as x-rays. While x-ray beams are used to deliver high doses of radiation to targets that might be deep within the body electron beams are used to deposit their energy near the surface, the depth of penetration being proportional to the electron energy. Accelerators with an electron facility can therefore be operated over a range of energies. In *The Physics of Electron Beam Therapy*, Klevenhagen (1983) covers this subject in depth but the following sections describe the production of the useful electron beams from the narrow beam that emerges from the accelerator and beam transport system.

### 6.5.1. Scattered and collimated electron beams

When operating in the electron beam mode, most dual-purpose linear accelerators use metal scattering foils to widen the pencil beam of electrons emerging from the thin window of the accelerator vacuum system The simplest arrangement is illustrated in figure 6.12. The x-ray target is removed from the beam and replaced by a thin window to allow the electron beam to emerge from the vacuum system. The electron beam is scattered by a thin foil which replaces the x-ray flattening filter (which if not removed would stop the electrons and become a rather inefficient x-ray target!) and is finally collimated to provide the required field size at the surface of the patient. Operation of an accelerator in the electron therapy mode imposes additional requirements on the design of other components in the treatment head. The mirror in the optical system has to allow passage of the different-energy electron beams. This requirement can be met by using tightly stretched aluminized plastic foil. Also the beam monitoring ionization chamber has to be sufficiently transparent to high-energy electrons, which is a more demanding requirement and will be discussed in chapter 7.

Electrons undergo significant scattering in the air between the scattering foil and the patient to be treated so, to produce a well defined beam, collimation has to be as near to the treatment surface as possible. The x-ray beam collimators, inside the treatment head, are not adequate for this purpose. Furthermore high-atomic-number materials, while appropriate to collimate the x-ray beam, would generate unwanted x-rays when the electrons are scattered. It follows that the electron beam should ideally be collimated by low-atomic-number materials such as aluminium. The aluminium thickness does not need to be as great as the electron range in aluminium, as the primary electrons strike the walls of the applicator at glancing angles. Empirically it is found that the aluminium thickness required is typically one-third of the electron range. In order for the operator to see the region to be treated, the applicator may stop a few centimetres from the surface, or may terminate in an end frame, stood off on pillars from the main part of the applicator. The end frame can serve the additional purpose of being a mounting for lead (or low-melting-point alloy) 'cut-outs' to provide irregularly shaped fields.

Foils are chosen to give the best compromise between scattering, energy loss and bremsstralung production. Ideally there would be wide-angle scattering and no energy loss or bremsstralung production but this is not possible. In practice copper foils are chosen and the thickness selected empirically for each energy so each beam adequately fills the largest field size that will be used. As shown in figure 6.12(b) the beam scattered by the copper foil has an approximately Gaussian profile: it is clearly not possible to produce a perfectly uniform beam. Improvements in uniformity can be achieved by making use of adventitious scattered electrons from the applicator walls. Although these are of slightly reduced energy, they add to intensity around the periphery of the beam and significantly increase the uniformity for large beams. The amount of applicator scatter can be adjusted,

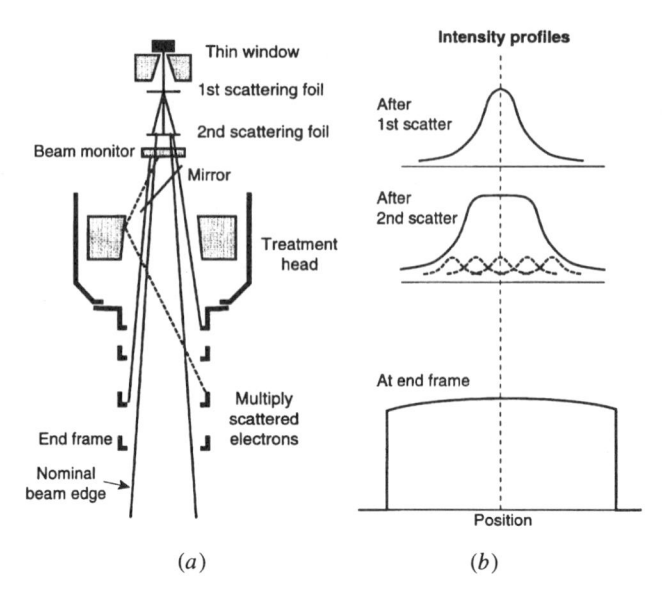

**Figure 6.13.** (a) An electron beam with dual scattering foils and open collimator. (b) The distribution of electrons scattered from the foils and collimator.

and uniformity optimized, by setting the x-ray collimators to control the number of electrons striking the walls.

An improved, but more complicated, scattering system is shown in figure 6.13. Here a second scattering foil is placed in the beam and each point irradiated by electrons scattered from the first foil becomes a source of scattered electrons. At the second foil some of the electrons at the periphery will be scattered back towards the centre of the beam and those near the centre will continue to be scattered outwards. The overall effect is to create a much more uniform beam than can be produced with a single foil and to reduce the need to inject lower-energy scattered electrons from the collimator. Figure 6.13 also shows an alternative collimator which can be used if collimator scatter is not required. It is an open structure consisting of a series of spaced 'picture frame' diaphragms. Unfortunately although this type of applicator appears lighter than the tubular structure it is not as the diaphragm thickness must be equal to, or greater than, the range of the highest-energy electron beam that will be used.

### 6.5.2. Scanned electron beams

The scattering process reduces the beam energy, increases the width of the energy spectrum and introduces x-ray contamination into the electron beam. The alternative, implemented on a few linear accelerators (and particularly on high-energy microtrons) is to extract the pencil beam from the beam transport system and use a magnet deflection system to scan the beam across the required treatment field. There are many possible scanning patterns but conceptually the simplest is the raster scan produced by applying saw tooth waveforms of different frequencies to two orthogonal pairs of deflection coils. This is similar to the scanning system in a TV tube or VDU but the scanning frequency and spatial resolution requirements are much less demanding. Although the beam is only a few millimetres in diameter as it emerges from the vacuum system it will, as has been previously stated, be scattered by the air in the treatment head. As a result the scanned beam is quite diffuse when it reaches the patient so a fine raster would not significantly improve control of beam uniformity or shape. (Scattering can be minimized by filling the head with helium, which has a very low scattering cross section, but this technique has only been applied in a few, very specialized, machines.) Secondly in view of the pulsed nature of the electron beam it is desirable to keep the raster frequencies low in comparison to the pulse repetition rate (PRF) of the accelerator. Typically the PRF is 250 Hz and the raster line frequency is 4 Hz, thus each line in the scanning pattern consists of approximately sixty pulses. The beam from each pulse is sufficiently diffuse that successive pulses merge into a continuous uniform dose distribution: this would not be the case if the ratio of the PRF and raster frequency were too low.

## 6.6. MULTIPURPOSE TREATMENT MACHINES

It should be clear from the previous sections that many of the components in the treatment head are specifically required and their detailed characteristics specifically tailored to the type and energy of radiation that is being produced. For example, the x-ray target is required for x-ray production but not for electron beam production, the scattering foil required for low-energy electrons is thinner than that required for higher-energy electrons etc. Therefore the treatment head for a multipurpose machine, which may produce x-rays at one or more energy and electrons at a range of energies, is considerably more complex than that for a single-mode machine. The positions of all the components that need to be selected for particular modes of operation are remotely controlled and are usually operated electro-mechanically. There is a further requirement for monitoring the position of each component so that interlocks can be provided to prevent operation with inappropriate combinations of components. These interlocks are also interconnected with the other accelerator systems including those for RF power control beam steering etc. The precision of these electro-mechanical systems must be extremely high as a small displacements can introduce major changes to beam intensity and uniformity.

The target and electron window which have to be interchanged between the x-ray and electron mode, are often an integral part of the vacuum system.

The shift mechanism requires a relatively small movement which can be achieved by use of a bellows, which can be compressed or extended without affecting the integrity of the vacuum system. Flattening filters and scattering foil can be mounted on a carousel, which is rotated to select the required filter or foil. If dual-foil scattering is used then two carousels can be used. The final movable component that may be fitted, apart from the collimators, is the universal wedge, which is mounted on a linear bearing and moves in the direction normal to the wedge gradient to avoid the need for extreme precision in positioning.

## 6.7. SCALES AND DISPLAY OF THE TREATMENT HEAD SETTINGS

Many of the settings of the treatment head are inherently digital: these include the position of the target and the position of each filter or foil; they are either 'in' or 'out'. Indication of the status of these settings are not usually local but they are of course monitored elsewhere. Other settings which are inherently analogue, or continuously variable, include the position of the collimators, defining the field size, and the angular position of the head rotated on its main bearing, which defines the beam orientation and most importantly the direction of the wedge when it is in use. These analogue parameters must be measured and displayed locally as an aid to setting up the treatment fields for each patient. Mechanical scales have, for the most part, been replaced by electronic measurement and display but where appropriate they can be very useful. They can be used as an accurate reference for the calibration of electronic readouts and do not suffer some of the limitations inherent in digital display: quantization errors and sometimes relatively slow refresh rates. Most machines include transducers on each of these variable items so that the position can be displayed electronically in digital form either on a VDU or on LED or LCD displays. The resolution required is 1 mm for linear parameters, such as collimator positions, and 1° for angular parameters such as the head rotation. (Note that, for the purposes of calibration, it is often desirable to increase the resolution by a factor of ten but not to display the least significant digit in routine operation.) Electronic measurement of treatment head settings facilitates their display and the same signals are used as input to automatic control systems and treatment verification systems, discussed in chapters 10 and 11.

## 6.8. CIRCUITS CONNECTED TO THE TREATMENT HEAD

The treatment head contains a large number of electrical and electronic components. These include the optical beam source, other optical beam direction devices, the drive motors of the beam defining collimators, the removable x-ray target and the carousel carrying the flattening filters and electron scatterers, which all need power supplies to be brought into the rotatable part of the treatment head. In addition, the treatment head contains transducers and monitors which provide signals to be taken to other parts of the accelerator's control system. Connection is normally made using a roll-up (retractile) multiway cable or cables which allow rotation of the system over about 370° and as more complicated functions are added consideration is given to multiplexing of signals to minimize the number of separate conductors required.

# CHAPTER 7

# THE DOSE MONITORING AND CONTROL SYSTEM

Small changes in the operating conditions, performance and even the ambient conditions of a linear accelerator can have significant effects on the radiation beams produced. It is therefore necessary to monitor the output continously to control the radiation beams in terms of energy and dose distributions and absorbed dose delivered to the patient.

## 7.1. THE BEAM MONITOR

The beam monitor consists of a set of transmission ionization chambers which monitor the whole cross sectional area of the radiation beam after it has passed the beam flattening filter or scattering foil. The three basic requirements for the monitor are

(i) that it should be 'thin',
(ii) that the sensitivity should be independent of ambient temperature and pressure and
(iii) that the ionization chambers should be operated under saturation conditions.

The monitor chambers must be 'thin' in relation to the radiation being monitored so that they cause minimal perturbation of the treatment beam. This condition can be satisfied for x-ray beams by constructing both the plates of parallel plate ionization chambers and the walls of their container from aluminium foil, as shown in figure 7.1. The ionization chamber consists of multiple parallel plates, each typically 0.1 mm thick. The parallel plate electrodes form a set of ionization chambers, which serve as detectors for two independent dose monitoring systems and for the dose rate monitor and include a segmented plate system allowing the dose rates in different parts of the beam to be compared.

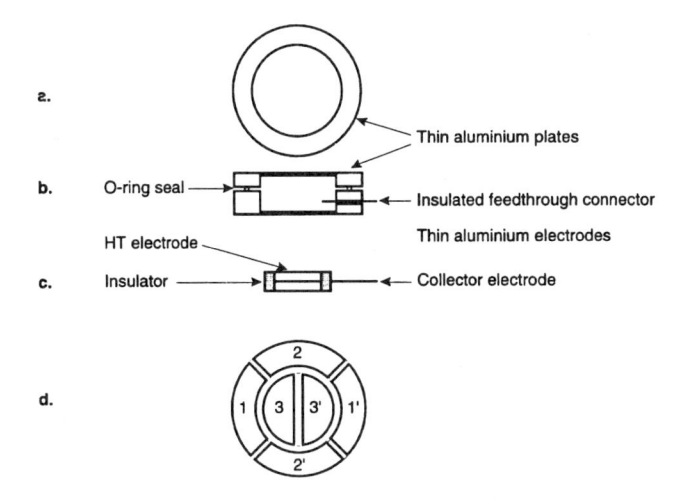

**Figure 7.1.** *Some details of beam monitor ionization chambers: (a) a circular section through a monitor; (b) a section normal to (a) illustrating the arrangement to make the system gas tight; (c) a multiplate arrangement; (d) a sectored plate to sample different regions of the radiation field.*

This is perhaps the simplest form of monitor but several variations in design have to be considered as the same ionization chambers are used for monitoring electron beams in multipurpose linear accelerators. Here it is necessary that the chamber should not cause excess scattering, production of x-ray contamination, broadening of the electron energy spectrum or generation of low-energy electrons. To minimize thickness, as measured by the mass per unit area, the electrodes are formed by deposition of carbon or metal onto a thin substrate of mica or a plastic foil (such as Melinex or Kapton).

The signal from an ionization chamber depends on the mass of gas between its electrodes. So, to maintain constant sensitivity, it is necessary to ensure adequate mechanical stability of the electrodes and avoid, or correct for, variations of ambient temperature and pressure. The mass of gas between the electrodes in a sealed enclosure is independent of ambient temperature and pressure as long as the walls are suffiently rigid to resist pressure differences across them. Many sealed systems have been used successfully over periods of many years, particularly on machines employed exclusively for x-ray therapy, where the chamber walls can be quite thick. The system is made gas tight by the use of the O-ring seals shown, and by bringing out the electrical connections to the plates through glass or ceramic insulated feed through connectors, of the type normally used on vacuum equipment. The chamber is usually filled with air. Clearly nominally sealed chambers that become unsealed will exhibit changes in sensitivity: tests of the integrity of the gas tight seals are covered in chapter 14.

The first requirement, for thin electrodes and walls, is difficult to reconcile with the competing requirements of geometric rigidity and insensitivity to changes in temperature and pressure. Thin walls deflect, in much the same way as an aneroid barometer, in response to any pressure differential between the sealed enclosure and its surroundings. So, although the mass of gas in the enclosure remains constant, the mass of gas between each pair of plates forming the ionization chambers changes with ambient pressure and temperature. The alternative to a sealed beam monitor is to leave it open to the atmosphere and to compensate for the inevitable variations due to temperature and pressure. This can be achieved at the expense of additional transducers for measurement of these parameters. The pressure can be measured remotely but the temperature must be measured in, or very close to, the beam monitor, because its position in the treatment head can be subject to substantial variations in energy deposition (e.g. from the light beam projector) and therefore variations in temperature.

The monitor, mounted between the flattening filter and the collimators is subject to very high average dose rates, up to 40 Gy $min^{-1}$, and, particularly during the radiation output pulses, peak dose rates of maybe 40 kGy $min^{-1}$. Under these conditions the property of the beam which determines the ion collection efficiency is the dose per pulse rather than the average dose rate. The ionization caused by each pulse produces such a high charge density in the gas in the ionization chamber that the normal competition between charge recombination and charge collection is biased in favour of recombination. This problem is discussed quantitatively by Boag (1966) and in *ICRU Report* 34 (1982) on the limiting assumptions that the total ionization per pulse is produced instantaneously and that the charge has been collected or recombined before the next pulse arrives. The collection efficiency for a monitor system, i.e. the proportion of the total charge collected, is not in itself important, so long as it does not vary significantly with the dose per pulse. This implies that the collection efficiency at the highest dose rates should be about 99%. The result of this discussion is that rather high field strengths are required in the ionization chamber, and this requires small plate separation, of about 1 mm, and a polarizing voltage of about 300 V.

As has been mentioned earlier the beam monitor is in fact a set of ionization chambers, each utilized for one of the control functions that will be discussed in the following sections.

## 7.2. MONITORING AND CONTROL OF BEAM UNIFORMITY—X-RAYS

Beam uniformity depends on delivering the electron beam accurately onto the x-ray target so that the focal spot is aligned with the beam flattening filter. It also depends on the electrons being accelerated to the correct energy because the flattening filter shape is specifically tailored to a particular energy. Any variation in energy will therefore cause the beam to be under- or over-flattened relative to the required uniformity. The systems which control the energy, position and magnitude of the electron beam arriving at the x-ray target are described in chapter 5, while the x-ray beam flattening filter is described in chapter 6.

It is implicit in this discussion that beam stabilization is necessary but perhaps this requires some explanation. When the gantry is rotated, or even when the patient support system is moved, the path of the accelerating electron beam is changed with respect to any stray magnetic fields present. This arises in relation to the Earth's magnetic field, and to any magnetization of the steel components of the gantry, the patient support system and the building where the machine is housed. These changes will have a significant effect on the flight path of the accelerated electrons. This is more marked in higher-energy machines as the length of the electron flight path increases.

There may also be some mechanical deflection of the components, with respect to each other, when the system is moved: for example the beam bending magnet may move slightly with respect to the accelerating waveguide. When the cathode of the electron gun is replaced, its position with respect to the anode may change. Also, during the life of a cathode, its position could vary, in so far as the heated components may not have been perfectly annealed before assembly.

The main electrical parameters (e.g. the modulator voltage and the microwave frequency) which determine the energy and position of the accelerated electron beam are all stabilized, but even when they are working properly they are not perfectly stable.

All these factors then may cause slight variations in the properties of the x-ray or electron beams, both in time and when the system is moved, and the beam stabilization systems are designed to correct these variations, within specified limits.

Having established the need, this section now discusses how signals from the monitor are used to regulate the position of the focal spot and the energy of the electrons as they strike it. While such controls are applicable to all types of machine the example used here is that of an accelerator with a 90° bending system shown schematically in figure 7.2. The six ionization chamber segments used for monitoring beam uniformity are arranged in three pairs, each pair providing signals dependent mainly on a particular beam control item, such as a steering current or the beam energy. So in figure 7.2(b), the plan view of the segments, there are two pairs aligned with the gun–target axis and a single pair aligned with the transverse axis. With an energy dependent bending system changes in beam energy have the greatest effect on the outer part of the beam and are therefore monitored with segments 1 and 1'. The inner part of the beam is more sensitive to changes in the position of the beam as it enters the bending magnet and

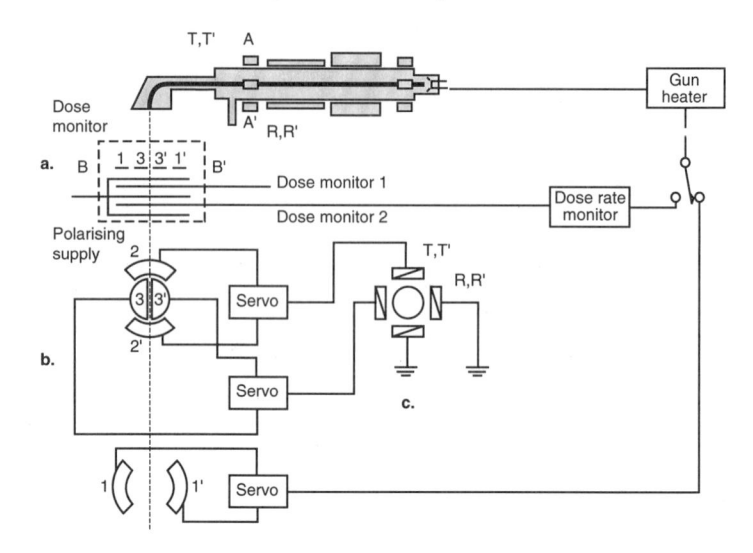

**Figure 7.2.** (a) A schematic diagram showing the relationship of dose monitoring ionization chambers to the beam steering system. (b) A section through plane BB' of the dose monitor chamber. (c) A section through plane AA' showing the beam steering coils.

these are monitored with segments 3 and 3'. In the transverse direction the energy has no effect and so the beam position in this direction is monitored by segments 2 and 2'.

### 7.2.1. Control of beam steering

The systems which control the position of the focal spot are the orthogonal pairs of beam steering coils at each end of the accelerating waveguide, the focusing coils, the beam bending magnet and the electron gun. The currents in the steering coils at the gun end of the guide, in the focus coils and in the beam bending magnetic coils are run at preset values from stabilized power supplies. The precise values of these currents are pre-determined, depending on the electron energy at which the accelerator is being operated. The currents in the beam steering coils R, R' and T, T', at the target end of the waveguide are variable and regulated by signals from the beam monitor, which is able to detect the effects of small changes in beam position. (The R and T labels refer to the radial and transverse deflection of the electron beam in the field of the bending magnet.)

The servo amplifiers shown as modules in figure 7.2(b) are shown in slightly more detail in figure 7.3. The outputs from the ionization chambers are conditioned to give equal signals when the beam is uniform. These signals form the input to a differential amplifier, or comparator, whose output is the error signal for the beam steering servo which is, of course, zero for a

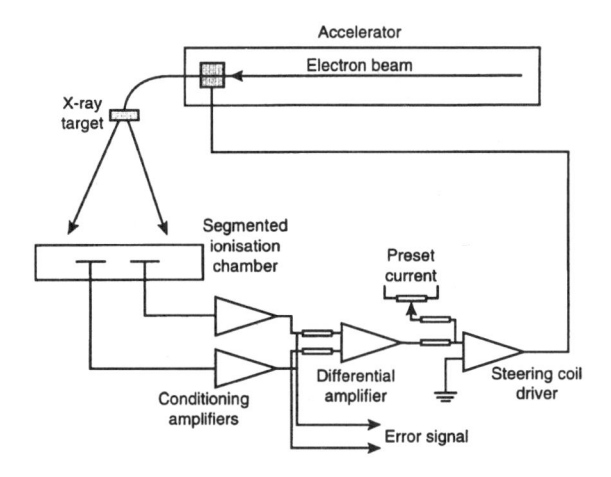

**Figure 7.3.** *A simplified schematic of the beam steering servo.*

uniform beam. The approximate beam steering current necessary to produce a uniform beam will have been set with the 'pre-set control' signal, which is added to the servo error signal to provide the regulated control signal to the steering coil power supply. Having established this feedback loop any variation of the beam position, caused by, say, the rotation of the accelerator in the Earth's magnetic field, will result in a differential output from the ionization chambers. The servo system will then correct the position by automatically and continuously adjusting the beam steering current about the level that gives a uniform beam.

The inputs to the comparator can also be monitored so that if the error signal exceeds an acceptable level corresponding to, say, 10% non-uniformity then an interlock will operate to stop irradiation. This is a requirement of IEC 601.2.1, the international standard specification for the safety of medical linear accelerators, which is discussed more fully in section 7.4.

### 7.2.2. Control of beam energy

As has been stated above, in an accelerator with a 90° bending system small changes in energy affect the symmetry of the outer parts of the x-ray beam and so the outer segments of the ionization chamber, 1 and 1', can be used to monitor and control the beam energy. The servo system is, in principle, identical to the beam steering servos. The most common way that electron energy is controlled is to exploit the beam loading characteristics of the accelerating waveguide. At the high beam current required for x-ray production the electron energy can be finely controlled by varying the electron beam current. If the energy is too low it can be increased by reducing the beam current and if it is too high it can be reduced by increasing the beam current. The difference signal from segments 1 and 1' is therefore used to regulate the electron beam current, via the gun heater supply (or in the case of a triode gun via the grid voltage), in a direction which will equalize the signals from segments 1 and 1'.

Although the signals from segments 1 and 3 are most sensitive to changes in energy and steering respectively there is some cross talk, so, if the energy stays constant and the beam position moves a degree of asymmetry will be measured by both pairs of segments. Similarly a beam whose position stays constant but which suffers a small change in energy will also produce asymmetry in both pairs of segments. There is therefore a potential for conflict between the two servos used to control the beam uniformity in the radial direction and for the conflict to result in oscillations. This can be avoided by ensuring that the time constant is very much greater for the gun servo than for the beam steering servo. It should also be noted that the input signals are pulses of current from the ionization chamber that are produced at the PRF of the microwave source, which may be as low as 50 Hz in some modes of operation. This puts an upper limit on the frequency response of the beam control servos.

### 7.2.3. *A further note on beam uniformity*

Control of beam steering and beam energy as described above is based on the principle of achieving beam symmetry, which is not synonymous with beam uniformity. It is possible to have a beam which is symmetrical but with either a concave or convex profile (i.e. with a hollow or hump in the centre of the beam). The most likely cause of this is that the beam energy is incorrect, and therefore not matched to the flattening filter, but that the combination of steering focusing and bending conditions have brought this 'incorrect' beam to the correct place on the target. Whatever their cause, it is necessary to be able to detect hollows and humps. The simplest way is to add additional segments to the beam monitor, one monitoring the periphery and the other monitoring the centre of the beam. The difference between these signals is then an additional measure of uniformity and can be used to activate the beam uniformity interlock. Alternatively the signals from the individual 'symmetry measuring' segments can be combined to produce a composite uniformity signal. In figure 7.2, if the 'inner symmetry' is logically represented by calculating the difference $3 - 3'$ and the outer symmetry by calculating the difference $1 - 1'$, the size of the hump in the centre of the beam can be calculated by $(3 + 3') - (1 + 1')$.

### 7.2.4. *Control of machines without bending or using achromatic bending*

In these machines the electron energy is not a critical factor in determining the position of the electron beam on the x-ray target. The electron beam position can then be regulated using the R and T steering coils of figure 7.2(a). By definition an achromatic bending system can bring a beam of incorrect energy to the correct position and this is also the case for an accelerator mounted in line with the treatment beam axis. The only effect of incorrect energy on the beam is to create a hump or hollow in the beam profile as discussed in the previous section. The electron beam energy can be controlled in these cases by using the signals monitoring the centre and periphery of the beam as input to a servo amplifier which then regulates the beam current as in the 90° bending system.

## 7.3. MONITORING AND CONTROL OF DOSE RATE AND BEAM UNIFORMITY—ELECTRONS

The electron beam current required in this mode is a factor of about a hundred less than that required in the x-ray mode, and at this level, changes in electron beam current do not significantly change the electron energy. Regulation of the electron beam current will therefore not stabilize the position, or the direction, of the electron beam at the exit window. However it is still necessary to regulate the beam current in order to stabilize the dose rate, which is critically dependent on the gun filament current (and hence temperature) in a diode gun or the grid voltage and phase in a triode gun.

For electron therapy, the beam flattening filter is replaced by the appropriate scattering foil, while the current in the beam bending magnet is set to the value for the electron energy required. The electron beam current can then be stabilized at the required value by pre-setting the gun heater current (for a directly heated cathode) to approximately the required level and then using the signal from the dose rate monitor section of the monitor chamber as the servo signal to provide fine control of the gun heater power supply, (figure 7.2(a)). Difference signals for segments 2, 2′ and 3, 3′ can then be used to regulate the currents in coils R and T to centralize the electron beam on the exit window.

The electron beam current can again be stabilized by the use of the signal from the dose rate monitor. If a triode electron gun is used this signal will control the grid voltage during the 'grid on' pulses.

## 7.4. DOSE AND DOSE RATE MONITORING

### 7.4.1. *Principles of a single-channel dose monitor*

Although it might be considered self-evident the principle of using a beam monitor, mounted in the treatment head, in order to measure and accurately

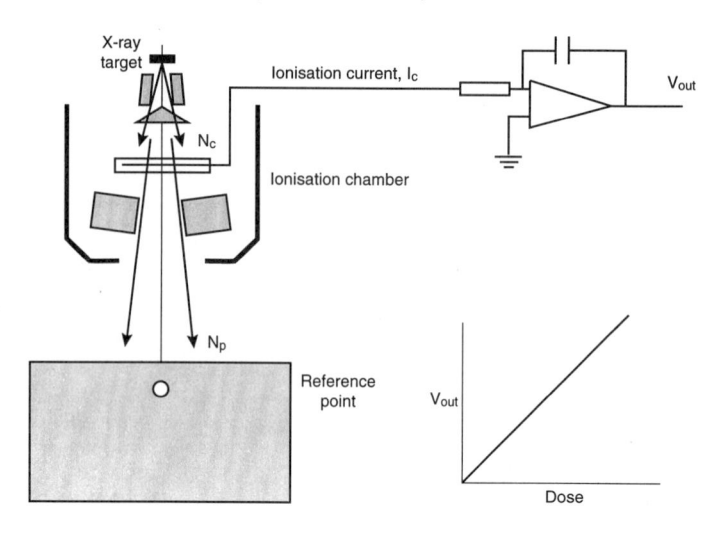

**Figure 7.4.** *The principle of measuring dose in a phantom using a monitor in the treatment head. The monitor ionization chamber current integrator is calibrated to measure dose at the reference point in the phantom.*

control dose delivered to a patient or phantom, some distance from the treatment head, is illustrated in figure 7.4.

The current, $I_c$, collected in the ionization chamber is proportional to the number of photons passing through the chamber, $N_c$, per unit time. Obviously the absorbed dose rate, $dD/dt$, at the reference point in the phantom is proportional to the number of photons incident on the phantom, $N_p$, per unit time. $N_p$ is clearly less than $N_c$ because the beam has been collimated and because of the inverse square law, but it is again proportional to $N_c$. It can be seen that the monitor ionization chamber current is proportional to the dose rate at the patient and, having established the constant of proportionality, the beam monitor current can be calibrated in terms of dose rate in the phantom for these reference conditions (i.e. for a given beam energy, field size, source to surface distance and depth in the phantom ).

If the current is proportional to dose rate, then the total charge collected during an irradiation will be proportional to the dose absorbed at the reference point. A signal proportional to the charge (the time integral of the current) can be obtained by a simple integration circuit as shown in figure 7.4. Assuming that capacitor, $C$, is discharged at time, $t = 0$ and the input current, $I_c$, is the current derived from the monitor ionization chamber then the output voltage, $V_o$, will be proportional to the dose delivered.

Although the first linear accelerators did not use solid state operational amplifiers their dosimeters were simple integrators of this type and the dose delivered was displayed on an analogue moving coil meter calibrated

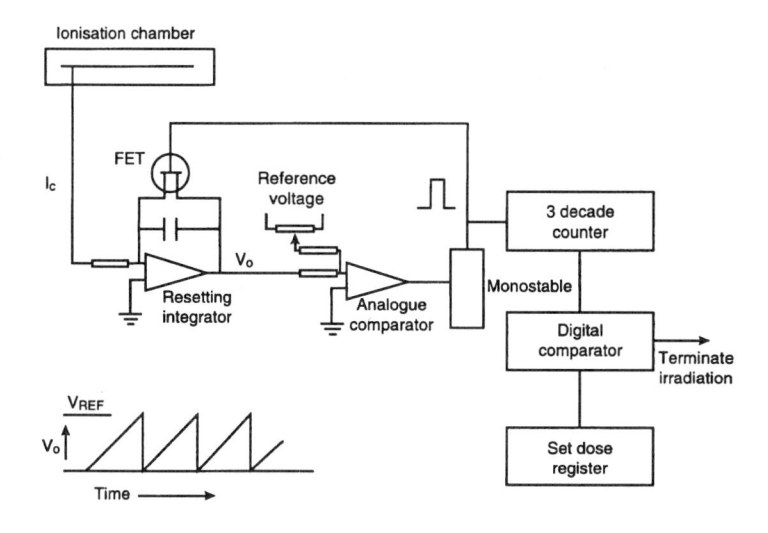

**Figure 7.5.** *A single-channel dose monitoring and control system.*

in terms of 'dose at the reference point'. Dose control was achieved by the operator switching the accelerator off as the meter needle reached the prescribed dose.

In order to automate the control of dose delivery and make use of digital electronics and more recently microprocessor and computer control systems it is necessary to convert the inherently analogue signal from the ionization chamber into pulses, each representing a pre-determined absorbed dose (at the reference point), which can be counted and compared with the prescribed dose for each irradiation. The absorbed dose corresponding to each pulse is then refered to as a 'monitor unit'.

The requirements for the electronic circuits of dosimetry systems are very stringent in terms of stability, reliability and safety. As a result the circuits are very complex and their detailed description are beyond the scope of this book. However, figure 7.5 illustrates the principles of a system based on the use of a resetting integrator and digital comparator to measure and control the delivery of dose.

The resetting integrator, which forms the analogue part of this circuit, is used to convert the ionization chamber current into monitor unit pulses. The ionization chamber current is integrated on the capacitor and the output voltage, $V_o$, is compared with a reference voltage, $V_{ref}$. When $V_o = V_{ref}$ the monostable is triggered and the output pulse turns on the FET, which, acting as a switch, discharges the capacitor allowing the process to start again. Each pulse, or monitor unit, corresponds to a constant charge and therefore an equivalent absorbed dose at the reference point. The amount of absorbed dose delivered by one monitor unit is 'calibrated' by adjustment of the reference voltage, $V_{ref}$. Usually $V_{ref}$ is set so that one monitor unit corresponds to 1 cGy at the reference point in the phantom for a given field size.

Having produced a stream of pulses the measurement and control of absorbed dose then becomes a problem of digital electronics, shown in block diagram form in the second part of figure 7.5. The pulses are counted using decade counters from which the number of monitor units accumulated can be displayed. The outputs from the counters are continuously compared with a register pre-set with the number of pulses required to be delivered. This register could be a second set of counters or most simply a set of binary coded decimal (BCD) switches.

The circuit described (which we stress is for illustration only and is not a practical design) constitutes a single-channel dosimetry module which includes components to

(i) produce electronic pulses, each representing one unit of dose (a monitor unit),
(ii) pre-set the prescribed number of monitor units,
(iii) count and display the number of pulses delivered during an irradiation
(iv) compare the number of monitor units delivered, $MU_d$, with the number prescribed, $MU_p$ and terminate exposure when they are equal (or when $MU_d > MU_p$).

Usually patient exposures are prescribed and monitored in terms of dose, and monitoring of dose rate is required mainly as an indication that the equipment is operating correctly. However, for moving beam therapy, the dose rate has to be stabilized at a pre-determined value, or used to control the rate of movement of the gantry. The dose rate signal may also be used to control the electron beam currents in the accelerating waveguide as mentioned earlier in this chapter. A dose rate signal can be obtained by monitoring the pulse stream with a frequency to voltage converter giving a voltage proportional to dose rate. Although this seems unecessary, as the signal taken directly from the ionization chamber is a more direct measure of dose rate, the more circuitous approach ensures that the same source of information is used for both functions. This ensures consistency in all control functions using both the dose and the dose rate signals, such as those required for dynamic therapy.

### 7.4.2. A dual-channel dose monitor

Because of the high dose rates available from linear accelerators a small increase in exposure time due to a monitor failure can result in a gross overdose to a patient. This is particularly true in the electron therapy mode, where in addition a change in electron gun conditions may increase the dose rate by a factor of hundreds. To deal with these possible changes a high level of redundancy is built into the dose and dose rate

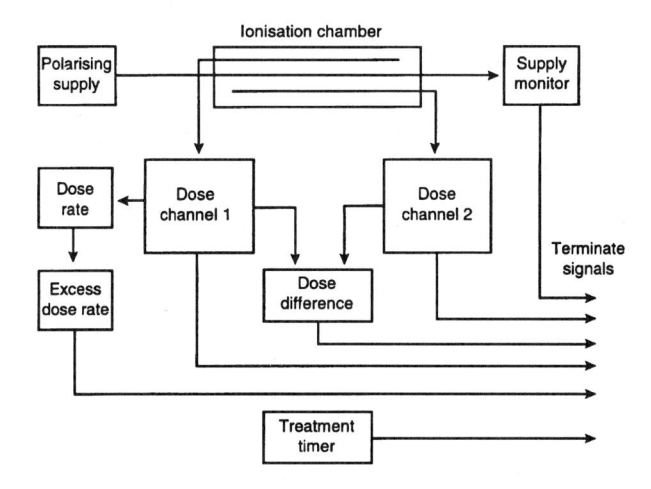

**Figure 7.6.** *A block schematic diagram of a dual dosimetry system to comply with IEC 601-2-1.*

monitoring systems under particular requirements for safety laid down by the International Electrotechnical Commission (IEC): IEC 601-2-1 (1981) and its amendements (1992). A second edition of this standard is due to be issued in 1997.

Figure 7.6 is a block schematic diagram of a dual dosimetry system which meets the requirments of the standard, as follows.

There are two independent ionization chambers, one providing the input signal to dose channel 1 the second to dose channel 2. If the ionization chambers are of the hermetically sealed type then they must be sealed in separate compartments so that air leakage to one will not affect the other.

In most cases the ionization chambers are permanently fixed in the radiation beam, but the standard allows for removable chambers on the condition that their correct positioning is interlocked.

The polarizing supply in this example is common to both chambers, so it is monitored continuously to ensure that if it falls to a level where either chamber would suffer a 5% change in response then irradiation would be terminated (or prevented from being started).

Each of the dose channels shown corresponds to the schematic diagram in figure 7.5. There are two displays associated with each channel, the first of the predetermined number of monitor units to be delivered and the second of the number actually delivered. Both these displays must be close together so that they can be observed by the operator. Many accelerators use video display terminals (VDTs) to present information to the operator and as these are vulnerable to failure it is necessary to either use two VDTs or have a conventional display (usually a battery powered LCD) as a back-up for at least one of the channels. This back-up facility also provides security against mains failure. After irradiation has been terminated by any means the dose displays must be maintained and before a new irradiation is initiated the dose counters should be tested and reset to zero. (In normal circumstances the irradiation will have been terminated by channel 1; the test of channel 2 is to simulate an input and check that it generates a terminate signal at the correct setting.)

The dose difference module continuously compares the output of the two dose channels so that possible errors are detected and acted upon before harm can be done. The level at which a dose difference triggers termination is equivalent to the smaller of 10% of the preset dose or 25 cGy (from IEC 1997 standard).

The system monitors the dose rate and terminates irradiation if this exceeds a level no more than twice the maximum expected. It is clearly necessary to ensure that the dose channels can function reliably at dose rates up to this trip level. Special consideration has to be given to the possibility of extreme dose rates that might arise and the most obvious cause is interlock failure, allowing operation of the acclerator in electron mode but with the beam current set to the x-ray mode level. In these circumstances an unacceptable dose could be delivered in a single RF pulse.

The final safeguard is the inclusion of a controlling timer which triggers termination in the event of undetected failure of both dose channels. This is set to a time which is not more than 120% of the expected time for treatment delivery.

The dose and dose rate monitoring system can only be calibrated to read dose and dose rate directly at a fixed distance from the radiation source (SSD) and at a given field size. It is convenient to adjust the sensitivity of the system so that each channel will deliver 1 cGy/monitor unit under these specified conditions. This can be done by adjusting the reference voltage for each of the comparators (shown in figure 7.5) of dose channels 1 and 2. For other field sizes, or SSD values, the relation between the readings given by the dose monitor circuits and the dose delivered is unambiguous, and suitable correction factors can be applied to relate dose monitor readings to dose in centigrays. Methods of calibrating the dose monitoring system and determining the necessary correction factors are discussed in chapter 13.

For a machine which can operate at a range of x-ray qualities and electron energies the relationship between the dose monitor reading and dose delivered at standard SSD and field size will depend on the radiation being used. It is again convenient to adjust the sensitivity of the dose monitor to deliver 1 cGy/monitor unit at a standard SSD and field size for each radiation used. This can be achieved by having a preset (and adjustable) family of reference voltage settings for dose channels 1 and 2 of figure 7.5, one setting for each of the radiations available. As an example, for an accelerator which can deliver two x-ray qualities and five different electron energies, amplifiers 1 and 2 would each need to have seven different settings.

These settings would then be automatically selected as one of the processes involved in setting up the machine to deliver the specified radiation for a particular treatment.

# CHAPTER 8

# BEAM DIRECTION AND BEAM SHAPING DEVICES

This chapter describes those devices which are used in setting up the radiotherapy machine to direct the radiation beam at the patient and if necessary to adapt its shape.

## 8.1. THE MAIN OPTICAL BEAM

This was mentioned in chapter 6, but will be discussed in further detail here. The basic system is shown in figure 6.1, where the optical beam is used to simulate the x-ray beam defined by the beam defining collimators. The system is set up so that the central axis of the x-ray beam, and the axis of rotation of the main bearing, are coincident, while the beam defining collimators are symmetric about this axis. The position of the light source (usually a quartz–halogen filament lamp) is adjustable to make the light beam defined by the beam defining collimators symmetric with respect to the axis of rotation. The shadow of a fiducial mark centred on the thin plastic window will then lie on the central axis of the x-ray beam. Alternatively the mark may be incorporated into the optical system nearer the light source.

*8.1.1. The relation of the x-ray field to the optical field*

Both the light beam and the x-ray beam have a finite-sized penumbra, and these cannot be exactly matched. Moreover the penumbra of the x-ray beam is not the same for all field sizes. For this reason the field size, as indicated by the light field, cannot agree with the size of the x-ray field to better than a few millimetres over the full range of available field sizes.

At this point it may be useful to consider the concept of the field size in more detail. Figure 8.1 shows the dose distribution across the x-ray field along a line perpendicular to the central axis. Figure 8.1(a) shows this in an idealized form, where the dose is perfectly uniform inside the field and falls

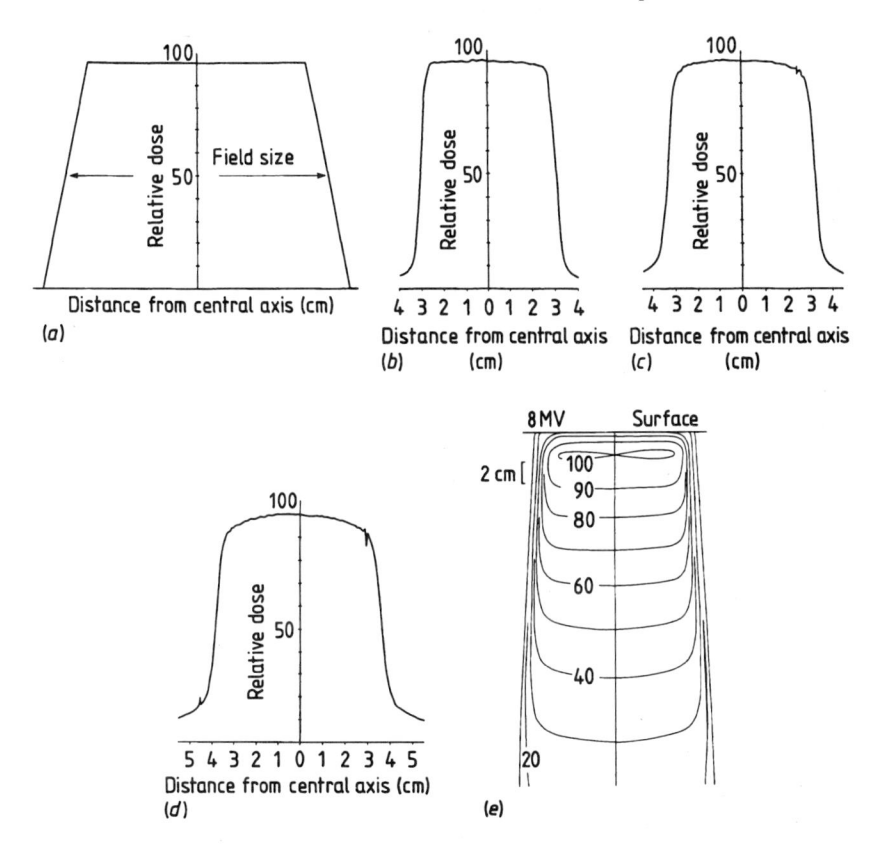

**Figure 8.1.** *(a) An idealized dose distribution across an x-ray field (perpendicular to the central axis). (b) The measured dose distribution at 2 cm deep. (c) The measured dose distribution at 12 cm deep. (d) The measured dose distribution 24 cm deep in a water phantom for 8 MV x-rays at 100 cm SSD. The field size is defined at 2 cm, the depth for maximum dose. (e) An isodose chart for 8 MV x-rays, at 100 cm SSD, corresponding to dose distributions (b)–(d).*

away over a finite distance as shown. Even in this idealized form it is clear that the field size must be specified at a particular dose level, usually 50% of the dose on the central axis, and at a particular distance from the source, usually at the isocentre (for machines that are isocentrically mounted). The fall-off in dose at the beam edge is not strictly a penumbra, although this term is normally used. In a medium the precise shape of the curve at the beam edge is determined by photons and electrons scattered from the main beam, by oblique penetration through the beam defining system, as well as by the size and intensity distribution of the radiation source (the focal spot). Figure 8.1(b)–(d) shows a number of measured beam profiles at different depths in a water phantom. These were measured with a small-diameter ionization chamber. It can be seen that the distance between the 50% dose levels increases with depth as does the width of the penumbra. In terms of specifying the field size necessary to cover the planned target volume (PTV) it could be argued that the 80% or even the 90% field width would be more appropriate than the 50% width. However field sizes based on these definitions do not scale accurately with depth or with SSD as well as the 50% field size does. It would therefore be even more difficult to ensure acceptable congruence of the light and x-ray fields and linearity of the x-ray field with indicated setting if any field size definition other than the 50% definition were chosen. The indicated setting is derived from the positions of the collimators within the x-ray head, which are measured with one or more potentiometers or shaft encoders.

The International Commission for Radiological Units and measurements (ICRU 1976) defines the distance between 50% dose levels as the field size. It is measured either at the depth of peak dose at the normal treatment distance or at the isocentre. There is some ambiguity in this definition, particularly at higher energies where the depth of the region of peak dose is several centimetres and the discrepancy between the two definitions will be several per cent. The preferred definition is the 50% width at the isocentre and at a depth of the peak dose. Thus, for an 8 MV machine with an SAD of 100 cm the standard measurement conditions for field size are 98 cm SSD and 2 cm deep.

The discussion above has assumed that the collimators are symmetrical as defined in chapter 6. For independent collimators the position of each of the 50% points relative to the axis is a more complete description of the field size than the distance between the two 50% points.

It can be seen that the relationships between the distance between the beam defining collimators, the scale which gives the 'collimator setting', the x-ray field size and the field indicated by the optical beam are not simple. The collimator scale setting can be calibrated to give x-ray field size as defined at one SSD, normally (SAD $-d_{max}$). The scale setting will have an unambiguous relation to any data, such as that in figure 8.1(b)–(d) and the isodose chart derived from it (figure 8.1(e)).

If the collimator setting is related linearly to the distance between the collimators, and this is usually the case, the relationship between the collimator setting and field size is not quite linear. This is because of the transport of energy out of the beam causing the distance between the 50% points to be slightly wider than would be expected from simple geometric considerations. The discrepancy becomes greater for larger fields; the 'penumbra' widens because of the effect of additional scatter.

Although the width of the optical field can only be measured to an accuracy of about ±2 mm because it also has a finite penumbra, the relationship between the indicated and the optical field size is linear.

The x-ray field size, the optical field size and the collimator setting may differ by several millimetres, up to 5 mm for the largest fields. This is recognised in the IEC particular requirements for performance. The guide to functional performance values, IEC 977 (1989b), allows 2 mm or 1% discrepancies between the light field and indicated size, 3 mm or 1.5% between x-ray field and indicated size, for field sizes up to 20 cm. The allowed discrepancies increase to 3 and 5% for light and x-ray fields greater than 20 cm wide. Further discussion of tolerances is given in chapters 12 and 14.

The outcome of this discussion is that the optical field does not give a sufficiently accurate indication of field size for small-field treatments. Field size can be accurately specified only in terms of collimator scale settings and the main function of the optical beam is to indicate the orientation of the x-ray beam, with respect to its central axis. For large-field treatment, for example regional or mantle treatments, the error between the optical beam size and the x-ray beam size is unlikely to be clinically significant.

One of the main advantages of linear accelerators over telecobalt machines is that they give very sharply defined x-ray beams. If the advantages from this for small-field treatments are to be realized, then the discussion in the preceding paragraphs is important.

If measured depth dose data and isodose charts are correctly labelled in terms of collimator scale settings, and if treatment prescriptions are also given in terms of this quantity, then the ambiguities and inaccuracies discussed in relation to optical and x-ray field sizes do not arise in the realization of a treatment plan as it is delivered to a patient. While this is true in terms of systematic errors that would arise if it were assumed that all measures of field size were identical, it must be remembered that random errors will arise from variations in field size between data collection, carried out during commissioning and checked periodically, and treatment delivery. Random errors on well maintained equipment should be less than $\pm 2$ mm.

### 8.1.2. *The relation of the electron field to the optical field*

The use of electron beam applicators is discussed in chapter 6. Figure 8.2(a) shows the cross field dose distribution provided by an electron beam applicator, while 8.2(b) shows the corresponding isodose chart. It can be seen that for electron beams the field definition varies significantly with depth, compared to that for x-ray beams where divergence is approximately geometric. The 50% points near the surface are reasonably well defined and used as a measure of field size but at greater depths electron scattering is added to the divergence, which is then not geometric.

Isodose charts for electron beams can still be unambiguously related to specified electron applicators.

Field size for electron beams is usually defined by the size of the end frame on the electron applicator (see figure 6.9). Where the end of the

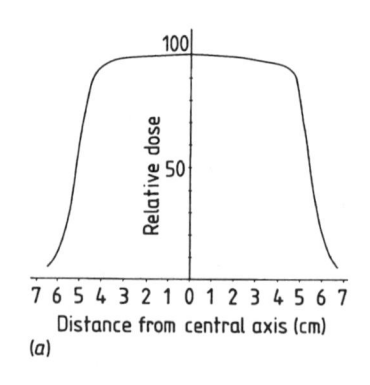

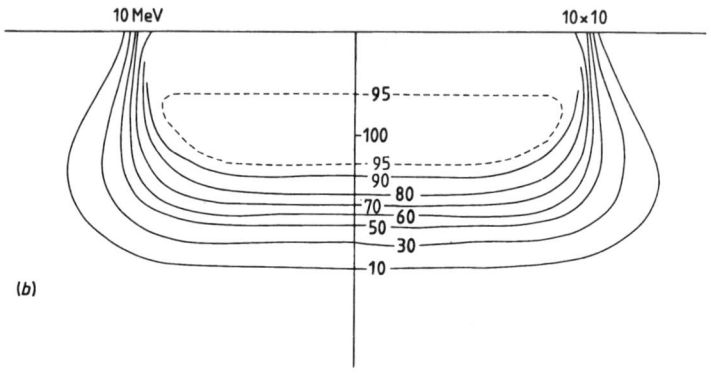

**Figure 8.2.** (a) The dose distribution across a 10 MeV electron beam 2 cm deep in a water phantom. (b) The isodose chart for a 10 MeV electron beam. The field size is 10 × 10 cm².

applicator is withdrawn from the patient's skin by a few centimetres, the optical beam will give a suitably accurate indication of field size.

## 8.2. MECHANICAL POINTERS

Mechanical front and back pointers are mounted on the accessory ring, as shown in figure 8.3, but not necessarily opposite each other. The pointers themselves are carried on rigid, usually tubular, members which are mounted on quick-release brackets on the accessory ring. Both front and back pointers are free to slide along the central axis, and are attached so that they can be easily pushed off if they come into contact with a patient. This is most easily achieved by use of steel pointers held on a permanent magnet. The front pointer carries a scale which relates to a reference mark on the mounting. The purpose of the scale is to measure the SSD and thus the scale reading is set to 100 when the tip of the front pointer is on the axis of rotation of the gantry (assuming an SAD of 100 cm). The magnets, or other mechanism,

*Optical pointers* 113

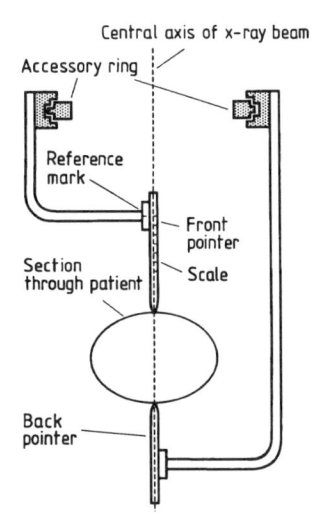

**Figure 8.3.** *Mechanical front and back pointers mounted on an accessory ring.*

holding the front and back pointers need to be adjustable so that the pointers may be aligned with the axis of rotation of the treatment head. In this way the front pointer can serve the dual purpose of measuring the SSD and indicating the position of the collimator axis at all SSDs within its range.

Mechanical back pointers are used less commonly and to achieve the necessary precision have to be very carefully engineered. Although the pointers themselves can be lightweight the mounting brackets are by necessity much longer than those for front pointers. Maintaining adequate rigidity of such a long lever is very difficult as a small mechanical misalignment at the mounting point on the treatment head can cause a large displacement of the indicated beam exit position. However, with careful maintenance and use, mechanical front and back pointers show the entrance and exit points of the central ray of the radiation beam on a patient with sufficient accuracy.

## 8.3. OPTICAL POINTERS

Most of the tasks for which mechanical pointers are used can be carried out with optical devices including the main optical beam which, as has already been stated, provides an indication of the central axis of the beam. Additional optical devices need to be added to measure distance along the axis.

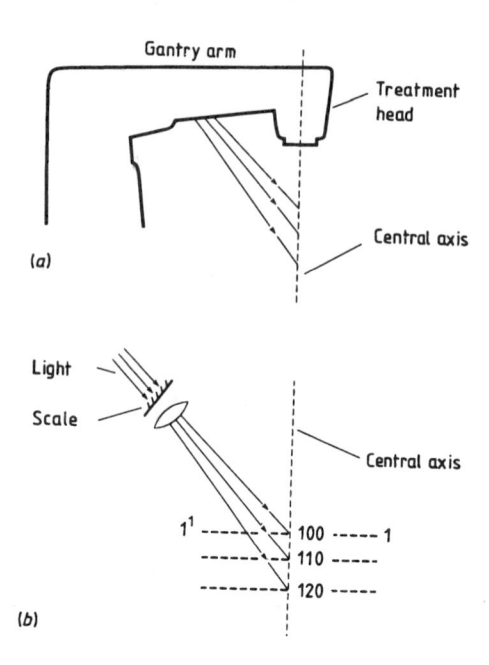

**Figure 8.4.** *The projected SSD scale: (a) basic arrangement; (b) some details of the optical system.*

### 8.3.1. The optical SSD scale

This is projected from a point on the gantry arm as shown in figure 8.4(a). More details of the optics are illustrated in figure 8.4(b). A scale is illuminated by a light source and condenser lens (not shown) and has its image projected by the lens. The scale is tilted with respect to the lens so that different points on the scale come into focus at increasing distances from the lens, and it can be arranged that the fiducial mark for each SSD is focused on the central axis at that SSD. The position of the central axis is indicated by the main optical beam. The image projected on the plane 1 1′ through the isocentre will then show the whole scale with the 100 cm mark in focus and in coincidence with the central mark projected by the main beam. Other points on the scale will be progressively out of focus with change in distance from the 100 cm point. At 110 cm the image will put the 110 mark on the scale in focus and on the central axis of the main beam.

This scale is particularly useful for setting up patients at SSDs beyond the isocentre, which is often necessary to provide treatment fields larger than those available at this distance, i.e. for field sizes greater than 40 cm.

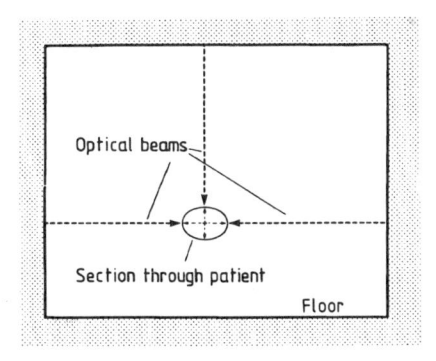

**Figure 8.5.** *Beam centring lights, mounted on the walls and ceiling of the treatment room, and converging on the isocentre.*

## 8.3.2. Axis lights

The optical systems mounted on the gantry and collimator are used to orientate the radiation beam relative to the patient. Before this is done it is often necessary or convenient to orientate the patient within the treatment room relative to the position of the isocentre and the direction of the rotation axes which define it. This alignment can be carried out using axis lights shown in figure 8.5. Two optical pointers, often referred to a side lights, are mounted on opposite walls of the treatment room to direct horizontal beams towards the isocentre, which is at the intersection of the axes of main rotational components of the treatment machine. A third optical pointer, a top light, is directed vertically downwards; all three converge at the isocentre. All three beams are in the vertical plane of rotation of the radiation source. In many installations the third beam is replaced or supplemented by another beam which projects a line along the gantry axis from a point vertically above the gantry axis of rotation. The vertical beam is obstructed by the machine when it is in the vertical position, in which case the same information is conveyed by the centre mark on the main optical beam. If the vertical beam is in the form of a cross, with lines about 1.5 m long, with one of these lines parallel to the axis of rotation of the gantry, this provides a useful reference for lining up the long axis of the patient.

The horizontal beams are useful for indicating that a recumbent patient is lying in the correct position. For instance in the case of isocentric treatments coincidence of the two horizontal beams with marks on opposite sides of the pelvis will ensure that the patient is lying 'flat', i.e. not rotated about the gantry axis. Coincidence of the three beams with three appropriate surface marks will place a desired point inside the patient at the isocentre.

Small movements and angular deflections of axis lights can result in large errors in alignment as the light beams are long optical levers. It is therefore very important to ensure that axis lights are mechanically stable and rigidly attached to a part of the room structure, ideally directly on the concrete walls. Most axis lights make use of low-power lasers with relatively simple additional optical components. The narrow beam emerging from the laser module is passed through a cylindical lens in order to produce a line. The intersection of two such lines produces a cross. The most common lasers are red, produced from helium–neon laser tubes or from laser diodes. Green lasers have also been used and are considered by some to provide more contrast on some types of skin.

### 8.3.3. Optical back pointers

Although back pointers are very useful, in order to establish the point at which the central axis of the radiation beam will exit from the patient, mechanical back pointers are somewhat cumbersome and prone to inaccuracy. An alternative is to use an optical back pointer formed by the intersection of a plane of light projected from a specific point on the gantry and the vertical plane of light from one of the side axis lights. This intersection occurs on the collimator axis of rotation which is the central axis of the beam. The point from which the optical back pointer is projected must be in the plane containing the collimator axis and the gantry axis. Careful consideration has to be given to the point from which the vertical line of the side light is projected so it is not obscured by the patient. This can be avoided if the line is projected from below the isocentre.

## 8.4. BEAM SHAPING DEVICES

The treatment head provides a continuous range of rectangular shaped x-ray fields, and electron applicators are also designed to provide rectangular fields.

### 8.4.1. X-ray fields

In the absence of an MLC, already described in chapter 6, it is necessary to use rather heavy blocks of high-density material to block off those parts of the rectangular field if an organ is to be shielded or a more complex shaped field is to be defined. These can be supported on a 'shadow tray' as shown in figure 8.6, suspended from the accessory ring of the treatment head. The main optical beam can then be used to indicate the shape of the field, subject to the penumbra limitations already discussed.

The shadow tray should be as far as possible from the surface of the patient to minimize the dose contribution from electrons scattered from the tray. Typically the surface dose increases from 25 to 50% but this increase is dependent on field size and distance from the tray to the skin surface. Also, by placing it as near the x-ray source as possible the size and weights

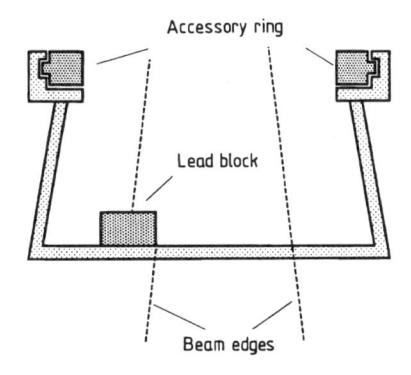

**Figure 8.6.** *A section through a shadow tray mounted on the accessory ring on the treatment head.*

of the block can be minimized. On the other hand, by placing the tray as near as possible to the patient, the penumbra on the beam edges defined by the blocks may be reduced. It is also necessary to allow sufficient space between the accessory ring and the top of the tray to get the blocks in and out. The compromise among these conflicting requirements is to place the shadow tray about 20 cm below the accessory ring.

The thickness of lead blocks required is not strongly dependent on x-ray quality in the present range (see figure 6.5). It is a compromise between achieving maximum attenuation of the blocked-off portions of the field and the fact that they have to be of a weight that can be lifted on and off the shadow tray. A thickness of between 5 and 7 cm of lead is usually considered necessary. 5 cm will reduce the radiation dose in the shadow of the block to about 5% of that in the absence of the block. This level is determined by attenuation in the block and by scatter into the shielded area of the field, and the precise value will depend on the field size and shape as well as the radiation quality involved. Increasing the thickness has only a marginal effect on the dose to the shielded region as, although the transmission is reduced, the scatter component remains constant and eventually becomes the limiting factor.

The Perspex tray, supported round the edges by a metal frame, needs to be about 10 mm thick to carry the weight of the lead blocks. It should be noted that irradiation of the tray over a long period will not only make the Perspex become yellow, but will also make it more brittle. For these reasons the Perspex needs to be changed regularly.

The shadow tray as described can only be used for a vertical beam. It can be made usable at other angles by fixing the blocks to the tray, either by drilling an array of holes in the Perspex through which the blocks can be screwed into place, or by sandwiching the blocks between two layers of Perspex. In practice a high proportion of the fields that require shaping are

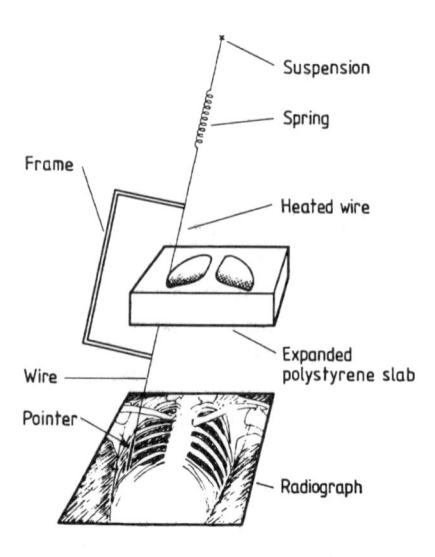

**Figure 8.7.** *A diagram of apparatus to cut projections of anatomical shapes out of an expanded polystyrene slab.*

vertical so that the simpler and more flexible system where the blocks are not fixed to the tray can often be used.

### 8.4.2. Moulded shadow blocks

Shadow blocks of complex shape can be cast in low-melting-point alloy using a technique illustrated in figure 8.7. The mould is cut in a block of expanded polystyrene of appropriate thickness using a hot wire. In figure 8.7 the wire is suspended at a point corresponding to the position of the x-ray source, while the expanded polystyrene block and the radiograph are at the distance of the shadow block and of the film when the radiograph was taken. If the pointer is then taken round the structure to be shielded, e.g. the lung shown in the radiograph of figure 8.7, the hot wire will cut out a section of the polystyrene of the required shape for the shadow block. This block will have tapered edges converging on the suspension, and will be of the correct shape to minimize penumbra. An alternative to the hot-wire cutter is to use a computer controlled machine to manufacture the polystyrene mould, in which case the shape can be defined in a computerized treatment planning system and automatically transferred to the block forming machine.

The polystyrene block can then be used as a mould to cast the appropriate shadow block in low-melting-point (eutectic) alloy (obtainable from Mining and Chemical Products). This material has a melting point of 60 °C so it can easily be melted and poured. In order to avoid over-heating the metal it is advisable to heat it in a water bath.

**Table 8.1.** *Data from the work of Berger and Seltzer (1964).*

| Electron energy (MeV) | Range in lead (cm) |
|---|---|
| 2 | 0.14 |
| 4 | 0.26 |
| 6 | 0.36 |
| 8 | 0.44 |
| 10 | 0.52 |
| 20 | 0.79 |
| 30 | 0.99 |
| 40 | 1.13 |

This technique makes it possible to cast shadow blocks to give fields of any desired size and shape. The blocks can then be used on the shadow tray and lined up with respect to the patient using the main optical beam. Alternatively, they can be fixed to a Perspex plate which attaches directly to the accessory ring, in which case they can be used for a beam at any angle.

The low-melting-point alloy has a density of about 0.9 that of lead, so shadow blocks 6–8 cm thick are regarded as adequate.

*8.4.3. Electron fields*

For more complex shaped electron fields, the alternatives are to use thin metal 'cut-outs' inside the end frame of the electron applicator or placed directly on the skin surface. At low energies it is convenient to use lead to define the field shape required. The thickness of lead required is determined by the range of the electrons involved in that metal, as given in table 8.1. At higher energies fabrication by casting low-melting-point alloy to the required shape is easier.

## 8.5. COMPENSATORS

The shadow tray can also be used to support tissue compensators. These are two-dimensional beam attenuators designed to compensate for 'missing tissue' and thereby overcome the effects of non-flat surfaces and internal inhomogeneities, which would otherwise give rise to non-uniform dose distributions. To a first approximation a compensator can be designed by measuring the depth of the missing tissue and calculating the missing attenuation over a two-dimensional grid across the beam. This map of attenuation is then projected back to the shadow tray and the compensator constructed of either thin sheets of lead built up like a contoured relief map or of a grid of aluminium blocks, typically projecting 1 cm × 1 cm areas of the beam to the isocentre. Alternatively a machine similar to the block cutter mentioned above can be used to cast compensators from low-melting-point alloy.

Compensation for missing attenuation can accurately correct the fluence of primary radiation but does not compensate completely for missing tissue. The scattered radiation generated in the compensator does not distribute energy in the same way as the scatter that would have been generated in the missing tissue. More accurate calculation of compensator thickness is therefore somewhat complicated and needs to take into account scatter either by approximation or ideally by iterative calculation based on an appropriate treatment planning algorithm, i.e. a pencil beam or convolution algorithm. It should be noted that although a compensator can improve dose uniformity it is designed for a particular plane in a patient and the cost of achieving uniformity in that plane can in some cases cause large dose variations elsewhere, for example at $D_{max}$.

As with beam shaping blocks there is an increase in the surface dose due to the secondary electrons scattered from the compensator and its mounting.

It can be seen that the design, manufacture and use of a compensator is somewhat inconvenient and time consuming. As a result they are not universally used. If the need for compensation is predominantly in one dimension a sufficiently good approximation can often be achieved with a wedge filter. Considerable efforts are being expended into the generation of compensated beams by the application of dynamic therapy with MLCs. This will be discussed in chapter 10.

## 8.6. GENERAL COMMENT

Linear accelerators can provide very precisely defined x-ray beams. To take full advantage of this, these beams need to be directed at a patient with comparable precision. The devices described in this chapter, plus the mechanical systems described in chapter 9, are all designed with this aim in mind.

# CHAPTER 9

# MECHANICAL SYSTEMS

During delivery of radiotherapy, a conscious patient is required to keep still. The best way to achieve this is to place the patient in a comfortable position and line up the radiotherapy generator with respect to the patient. The systems to be described in this chapter are designed to perform these functions.

## 9.1. THE ISOCENTRIC MOUNTING

The three basic arrangements for mounting the accelerating waveguide to allow the radiation beam from a linear accelerator to be directed at a patient were outlined in chapter 1. In clinical practice it takes much longer to set up the patient for treatment than to deliver the radiation dose. Therefore efficient utilization of the equipment requires that systems for lining up the radiation field with respect to the patient allow the setting up to be done in logical sequence, rather than by a process of successive approximations. The principles of the isocentric mounting used for this purpose were first described by Howard-Flanders and Newbery (1950) and are illustrated in figure 9.1.

The radiation source Y is mounted on the gantry so that it can be rotated about a horizontal axis X–X'. The radiation beam defined by the treatment head has a rotation axis Y–Y', which remains in a vertical plane, shown in figure 9.1(b). For all positions of the gantry the collimator rotation axis X–X' will cross the Y–Y' axis at the point I. The patient couch C is mounted on a turntable T–T' so that it can also rotate about a vertical axis through I. The couch can be raised and lowered on the pillar P, and moved longitudinally and laterally in the horizontal plane. The pillar P is placed off centre on the turntable to allow the gantry arm G–Y to pass under the couch when it is rotated. The axis of rotation of the source Y, the vertical axis of rotation of the couch and the central axis of the radiation beam all pass through a single point in space defined as the isocentre I.

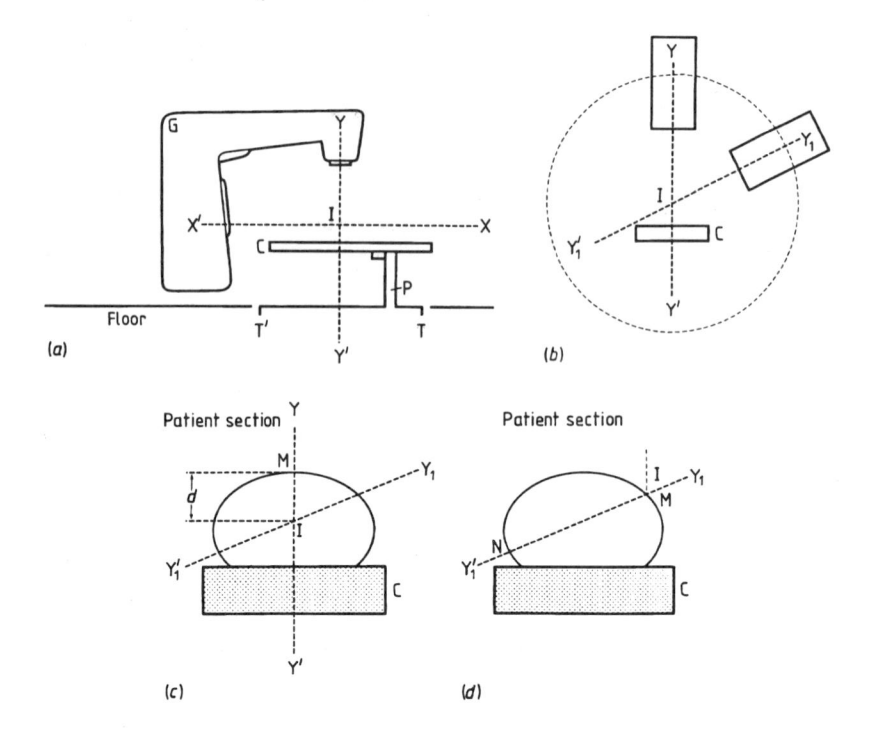

**Figure 9.1.** (a) A treatment unit: the horizontal line X–X' is the axis of rotation of the gantry: the vertical line Y–Y' is the axis of rotation of the collimators. (b) An outline in the plane through Y–Y', which is normal to X–X'. $Y_1$–$Y_1'$ is a second beam direction. (c) An illustration of an isocentric treatment. The two beams intersect at a depth d inside the patient. (d) An illustration of one beam of a fixed-SSD treatment. The isocentre is at the skin surface.

This system may be used in two alternative ways.

(i) By placing the centre of the volume to be treated at the isocentre: these treatments are often called isocentric treatments. The SAD is constant but the SSD is variable depending on the depth of the target.
(ii) By placing the entrance point of the radiation beam central axis at the isocentre: these treatments are often called fixed-SSD treatments as all beams will have the same SSD, which is equal to the SAD.

## 9.1.1. Treatment volume centred on the isocentre

This system is described in figure 9.1(c), which shows a section through a recumbent patient on the couch C. The minimum information required to direct a radiation beam at any required angle through the centre of the tumour is a skin mark M vertically above the tumour in this section, and the distance $d$ between the skin mark and the centre of the tumour. The gantry is rotated so that the central axis Y–Y' is set at the vertical position, and the vertical and horizontal movements of the couch are used to bring the mark M on the patient to the collimator axis at an SSD given by $(SAD - d)$ as indicated by one of the mechanical or optical systems discussed in chapter 8. This places the centre of the tumour at the isocentre, I. A radiation beam from any direction in the vertical plane can then be directed at the tumour by rotating the gantry through a specified angle. The central axis of the second beam is shown as $Y_1$–$Y'_1$ in figure 9.1(c). Note that the section through the patient does not have to be a right section. Any vertical section may be chosen by rotating the couch turntable about a vertical axis. This arrangement can be used to direct multiple fixed beams through the centre of the tumour, or for rotation therapy a single beam is rotated about the same centre. Because the couch rotation is isocentric it is not necessary to limit multiple beams to be co-planar although the use of non-co-planar beams complicates the treatment planning process.

*9.1.2. Beam entrance centred on the isocentre*

The alternative way of specifying the direction of the central axis of the radiation beam through a patient is to provide marks indicating the intended entrance and exit points of the central axis on the skin surface. These marks can then be used to set up a patient for treatment as in figure 9.1(d). This again shows a section through a recumbent patient on the couch C. The skin entrance point M can be brought to the isocentre I by using the vertical and horizontal movements of the couch. The exit point N of the central axis, indicated by the back pointer, can then be brought to the required point N on the patient by rotation of the gantry and the couch. The entrance point will not be moved in this process, as both these rotations are about I. The benefit of the isocentric mounting is that the positions of the entrance and exit points can be adjusted independently. The entrance positioned entirely by the linear movements and having done that the exit positioned entirely by rotational movements.

## 9.2. THE GANTRY

The accelerating waveguide, the focusing and steering coils, the treatment head and any necessary additional radiation shielding are mounted on the horizontal arm of the gantry which rotates about the horizontal axis X–X' of figure 9.1. It may be supported on a drum mounting or pendulum mounting as described in the next sections. The system as a whole will be subject to elastic deflections, and there will also be play in the bearings for the gantry rotation and for the rotation of the x-ray head. As a result of these

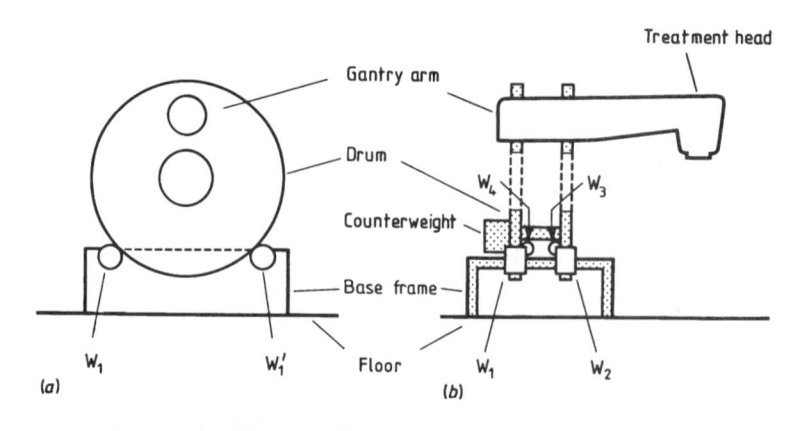

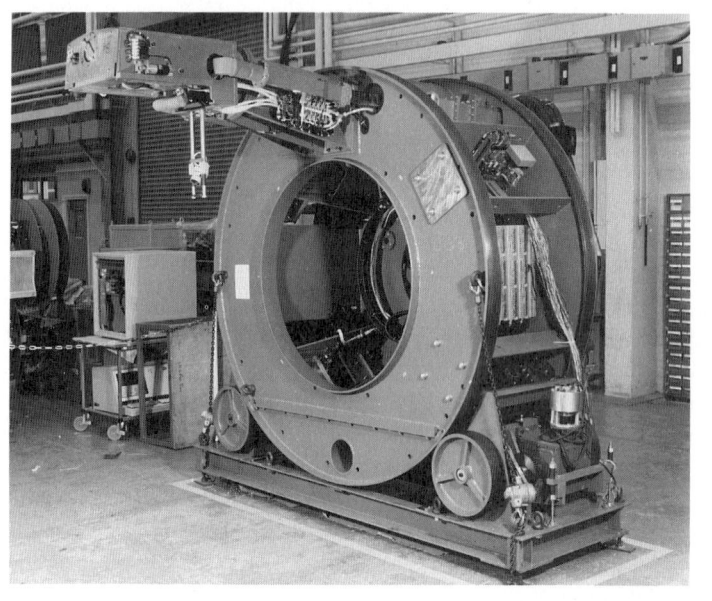

**Figure 9.2.** *An outline diagram of a drum mounted gantry. (a) A section normal to the rotation axis. (b) A section through the axis and beam arm. (c) A photograph of a partly assembled gantry. (Courtesy of Philips Medical Systems—Radiotherapy.)*

deflections and the play in the bearings the isocentre is not a unique point in space and will follow a complex path when the gantry is rotated through 360°. The system has to be designed such that the isocentre remains within a 4 mm diameter sphere.

*9.2.1. The drum mounting*

In this system the gantry arm is carried on a cylindrical or drum shaped structure, shown in section in figure 9.2(a) and (b). Figure 9.2(c) is a photograph of the structure during assembly. The outer rims of drum are turned to be concentric with each other and rest on the wheels shown. The diameter of the drum is mainly determined by the distance between the radiation source (the x-ray target, or electron beam window) and the centre of rotation, which in most units is 1 m. The height of the axis of rotation above floor level is sufficient to allow 360° rotation of the whole structure. This is usually between 1.2 and 1.4 m. The gantry arm does not have to be exactly horizontal. By tilting it upwards, it is possible to reduce the diameter of the drum, and still maintain an SAD of 1 m. In this case the beam bending magnet in the treatment head will have to bend the accelerated electron beam through neither exactly 90° nor 270°, as described in chapter 5, but through an appropriate angle to bring it on to the x-ray target in the direction in the vertical plane as shown in figure 9.1(a).

The horizontal arm may extend on both sides of the drum, depending on the length of the accelerating waveguide. The solid steel counterweight shown balances the system both about the axis of rotation and about the support wheels, to prevent it from tipping forward. Although simple in principle the design of these structures is very complex as they have to maintain sufficient rigidity at all angles. For example, although the system in figure 9.2(b) is balanced at the angle shown the drum will be subject to quite different twisting forces when it is rotated through 90°. The varying deflections arising from these varying twisting forces are the main reason for the isocentre not being a perfectly defined point.

The drum is rotated by driving one of the pairs of support wheels, $W_1$, the other pair, $W_2$, providing passive support. A further two pairs of guide wheels, $W_3$ and $W_4$, are supported off the base frame and rotate at horizontal axes. These wheels prevent the system drifting sideways on the main support wheels, i.e. parallel to the axis of rotation of the support wheels.

The base frame can, if necessary, be mounted in a shallow pit to minimize its height above finished floor level.

The drive wheels which rotate the system are powered by a servo controlled motor and gearbox whose speed is stabilized by a voltage fed back from a tachogenerator mounted on one of the drive shafts. This system allows a continuous range of drive speeds, from zero to the maximum available in either direction. It is varied either by a manually controlled input voltage or during rotation therapy by a voltage derived from the dose rate signal of the beam monitoring system. The drive system is self-braking, in the sense that as the motor control voltage is brought to zero the drive velocity will also come to zero. When the system is at rest, the fact that it is balanced to within a few kilograms enables the friction in the gear box to hold it in its rest position. The maximum time required for a full 360° rotation of the gantry is about 1–2 min.

The inside of the gantry drum provides ample space for those control circuits which have to be rigidly connected to the accelerating waveguide via the transmission waveguide system and also for the other elements providing voltage supplies to the accelerating waveguide and the electron gun circuits. A very large number of cables and some water pipes entering from fixed positions have to rotate with the gantry. In addition, for machines powered by klystrons the microwave power has to be fed onto the rotating structure through a rotating waveguide joint on the axis of rotation. The system cannot be continuously rotated because of the cables and water pipes: its range of motion is restricted to about 380° by direction sensitive fixed limit switches. After rotating through 380° the system has to be reversed.

The drum mounting provides a relatively rigid and stable system and has to be a very large and heavy structure, weighing altogether about 5 t.

### 9.2.2. The pendulum mounting

In this system the weight is carried on a vertical stand which is firmly fixed to a frame embedded in or bolted down onto the floor whilst the horizontal arm is carried on a main support which is free to swing as a pendulum (figure 9.3). The main support pendulum is mounted on the vertical stand via a slewing ring, a large-diameter thrust bearing which can support the moments about both the horizontal and vertical axes. The vertical stand and the main support are box structures. The gantry is rotated by a drive shaft passing along the central axis of the slewing ring, and is connected to the servo drive motor via a large-diameter pinion, or toothed wheel, which is driven from the main gearbox. The comments already made about speed control for the drum drive system also apply here. The rotating system may be counterweighted as indicated in figure 9.3(a), or by the use of a primary beam catcher as shown in figure 9.3(b). This carries a metal slab of sufficient dimensions to attenuate the primary beam dose rate by a factor of up to $10^3$ and reduces the required wall thickness for the treatment room (see chapter 13). It has the disadvantages of restricting access to operators when patients are being set up for treatment, and of limiting the distance at which the patient can be placed from the radiation source. Some machines are provided with a retractable beam catcher as a method of overcoming the second of these disadvantages. In this case the retraction mechanism must be controlled by a set of limit switches so that the catcher can only be withdrawn when the primary beam is pointing in directions where adequate attenuation is provided in the structure of the treatment room. For example, with a ground floor treatment room with no basement underneath, retraction of the beam catcher could be allowed when the radiation beam is directed towards the floor.

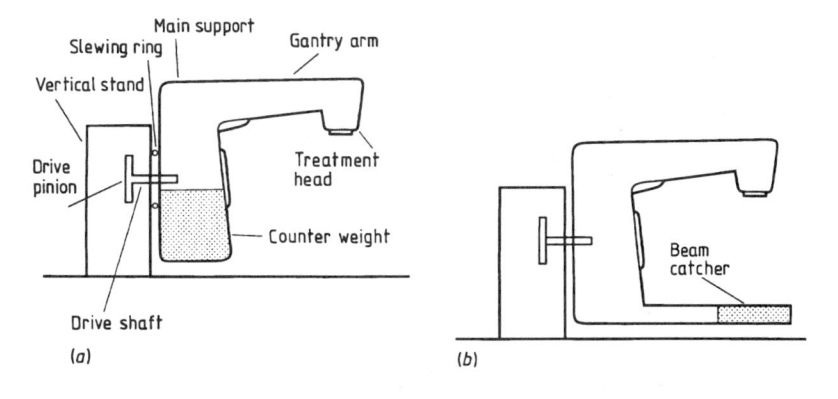

**Figure 9.3.** A gantry mounted on a slewing ring. (a) A system with a counterweight. (b) A system with a beam catcher. (c) A photograph of a partly assembled gantry. (Courtesy of Philips Medical Systems—Radiotherapy.) The base structure is below floor level when the system is installed.

Figure 9.3(c) shows a photograph of a linear accelerator on a slewing ring mounting with the covers removed. Again, with the pendulum mounting, a number of cables and flexible water pipes have to pass into the rotating structure and this limits the total angle of rotation as before.

### 9.2.3. Comparison of drum and pendulum mountings

Both types of mounting provide acceptable solutions to the problem of rotating the radiation source and beam defining system round a patient positioned at a fixed isocentre, and from a user's point of view there is little to choose between them. The drum mounting is often built into the treatment room so that the drum is behind a facia panel which forms part of a wall dividing the treatment room from the equipment space. In this case the drum does not need cosmetic covers and the components mounted on it are easily accessible for maintenance. This approach has also been used for some bigger pendulum mounted systems in which the fixed stand is located behind a facia panel in an equipment space. The alternative is to provide cosmetic covers for the drum or vertical stand and to install the machine in a single room. This is less convenient for maintenance but, aesthetically, there is a greater feeling of space in a single large room and this might be preferred by both operators and patients.

### 9.3. PATIENT SUPPORT SYSTEMS

The patient support system consists of a horizontal couch, or treatment table which can be moved to the required position by vertical, horizontal and rotational movements. In an isocentric mounting the main rotation axis of the couch is about the vertical axis through the isocentre. The range of vertical movement required is determined both by reasons of patient comfort and convenience and by the need to treat some patients at extended SSD in order to achieve field sizes greater than those available at the isocentre. Although many patients requiring radiotherapy are relatively fit the average patient is more than 60 years old and is likely to have limited mobility. It is therefore highly desirable to be able to lower the couch top down to a point where most patients will be able to sit and then lie on the couch without too much assistance. This lower limit of vertical movement also allows for extended SSD treatment with a beam of radiation directed vertically downwards. Extended SSD treatments are usually carried out by the use of a parallel opposed pair of beams, in order to produce a uniform dose distribution throughout the patient. In this case it is necessary either to turn the patient over, from a supine to prone position, or to lift the couch top to the required height above the isocentre and rotate the gantry to irradiate the patient with a beam of radiation directed vertically upwards. The second

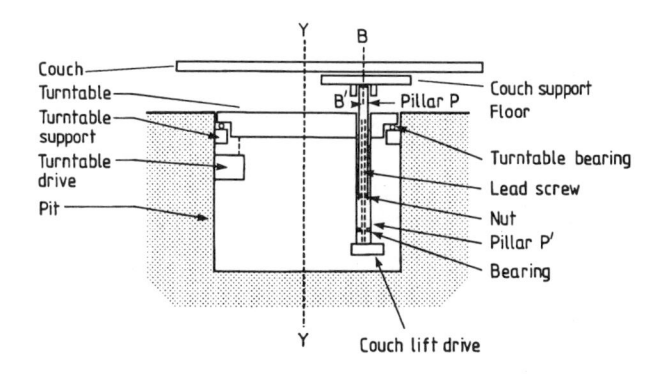

**Figure 9.4.** *A patient support system mounted in a deep pit.*

option avoids the possibility of significant movement of internal organs between the application of each beam and limits the handling of the patient. However it introduces the requirement for the patient support system to have a very long range of vertical movement.

There are two basic forms for the lifting mechanism, which can each provide the necessary range of vertical movement required for the couch top. These are illustrated in section in figure 9.4, where the lifting mechanism is mounted in a deep pit, and in figure 9.5, where the lifting mechanism is entirely above the floor.

### 9.3.1. Systems using a ram lifting mechanism

Figure 9.4 shows the system mounted in a deep pit. The turntable carrying the lifting ram is mounted on a bearing so that it can rotate about the vertical axis Y–Y'. This rotational movement is motor driven. The coupling between the fixed turntable drive motor and the turntable, indicated by a broken line, may be a belt, or a gear train. In this diagram, of an early design, the ram is mounted inside the large thrust bearing, whose diameter is about 1.5 m: however an alternative arrangement, where the ram is mounted outside a centrally supported smaller bearing, is now in use. The principle of operation remains the same. The cylindrical steel pillar P is supported on a lead screw inside the ram housing P' and can be raised and lowered by rotating the lead screw, the weight of the whole movable system being carried on the drive nut shown. Guide wheels or bushes between the inner and outer pillars P–P' prevent sideways play between the two surfaces. The ram and its housing have to be sufficiently stout to prevent significant deflection under the bending moment from the weight of the couch support and couch, and of a patient in the maximum off centre position. The lead screw is motor driven, as shown, the motor being servo controlled to give a continuously variable speed from zero up to that required to move the patient through

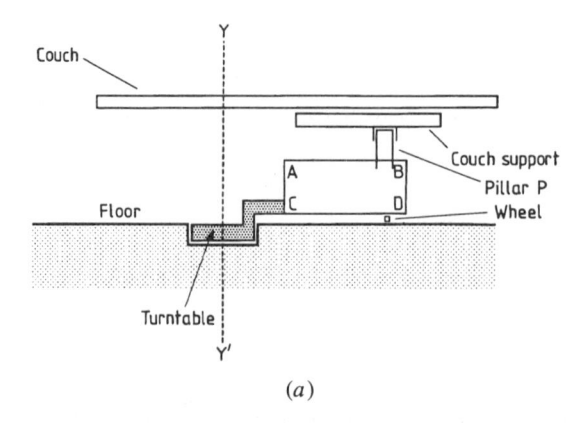

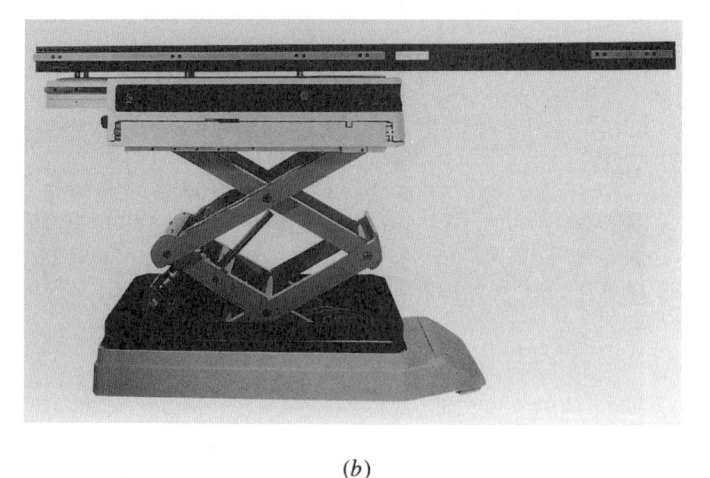

**Figure 9.5.** *A patient support system mounted above the floor: (a) a schematic diagram; (b) a photograph of a patient support system using a scissor lift. (Courtesy of Varian Oncology Systems.)*

the whole range of vertical movement in about 1 min. The lead screw nut and gearbox provide a non-reversible system so that the vertical position is stable when the drive motor is de-energized. Limit switches de-energize the motor when the nut approaches either end of the lead screw.

This system will then lower the couch support to just above floor level, allowing the maximum possible SSD for a radiation beam directed vertically downward. The patient may also be raised to any desired height above the isocentre by making the P–P' system sufficiently long. In other words the system will permit the application of the maximum possible field size to a patient, using vertically downward and vertically upward directed radiation beams. The deflection and bearing play for the whole system must be such that the centre of rotation for a patient, when the turntable is rotated, remains within a vertical cylinder 4 mm in diameter, whose long axis passes through the isocentre.

The advantage of this system is that it provides maximum vertical movement for the patient, and allows the application of very large fields to a recumbent patient from either above or below. The disadvantage is that it requires a deep pit which may be up to 2.5 m below floor level. The whole system weighs 1–1.5 tonnes.

### 9.3.2. Systems using a scissor lifting mechanism

A system requiring less space below floor level than the ram system is shown in figure 9.5. Here only the turntable and its bearing need to be below floor level. The couch lift mechanism is then carried in the box ABCD, which has to be placed off centre to allow the gantry arm to pass under the couch. The box, or pedestal, is rigidly fixed to the turntable and the weight of the couch, the couch support and the couch lift mechanism may all be cantilevered off the turntable and may be partly supported by a wheel running on the floor, as shown. The couch lift mechanism consists of a multistage scissor jack which is lifted by a lead screw and nut which open and close the mechanism. The bearings at each of the scissor joints must be of very high quality as any small deflections will be compounded and result in inaccuracy and instability in the lifting geometry. Scissor lifts with comparable travel to the ram lifts described above have been used successfully and have the advantage of not needing a pit and are more accessible for maintenance operations. The overall weight of the moving parts of each system (including the patient) will not be very different and therefore the power requirements will be the same. However, for the same rate of vertical travel, the lead screws will need to operate at quite different speeds. For the ram, the length of the lead screw is the same as the range of travel whereas for the scissor lift the length is much shorter and depends on the number of stages and the length of each arm.

The casing around a scissor lift system can be either a telescopic or concertina-like tube which has the same range of extension as the lifting mechanism.

### 9.3.3. The patient couch

The patient couch is mounted on a horizontal carriage allowing movement longitudinally and laterally. This is shown as the couch support in figures 9.4 and 9.5. The carriage includes the motorized mechanisms to drive the couch in each direction and clutches that can be released so that the couch can be moved manually in the same directions. With the couch 'floating' freely the patient can be moved quickly into approximately the correct position and the motor driven movements used to make final fine adjustments. The

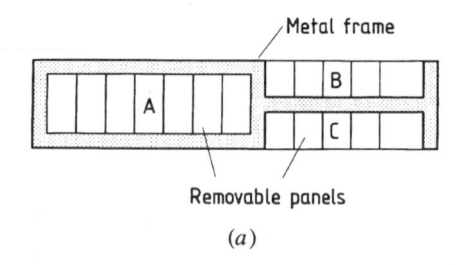

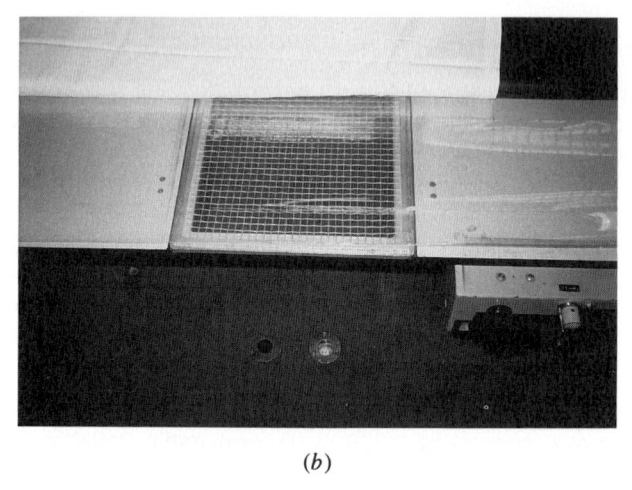

**Figure 9.6.** *(a) A diagram of the couch top. (b) A photograph of a 'tennis racket' support panel.*

friction in the motor driven movements also serves to hold the couch in the final position when the motors are de-energized.

The couch top consists of a metal frame into which removal panels are inserted and over which a thin, but strong, plastic film is stretched. The removable panels are to facilitate the irradiation of patients from all directions while maintaining a reasonable degree of skin sparing. The plastic film provides a continuous weight bearing surface to the couch when one or more panels are removed. The main frame of a couch top, shown in figure 9.6(a), is constructed with solid side members at one end so that when one of the panels, A, is removed there is free access from a beam directed vertically upwards. The other end of the frame has removable panels between a central spine and supported on lateral ribs. These are brought into line with the radiation beam by rotating the frame about its centre. Removal of a panel from section B or C exposes the posterior oblique surface of the patient to give minimum obstruction to the radiation field when, for example, irradiation with three equally spaced beams is employed.

The removable sections allow access to the entrance points on the skin surface, which is required both to be able to see the region at which the treatment beam is directed for alignment purposes, and to minimize the skin dose. Although the patient can be adequately supported on a couch with a 30 cm wide panel removed from any section, it feels very insecure to lose solid support from the hip or shoulder area. The conflicting aims of providing direct support in the area from which the panel has been removed and giving direct access to the patient through the couch can be satisfied by replacing the solid panel by one strung like a tennis racket, as shown in figure 9.6(b). An alternative to the tennis racket is to use a lightweight composite material formed by two thin sheets of carbon fibre, made rigid by bonding them onto a plastic foam panel. This maintains a reasonable degree of skin sparing but as it is not transparent it prevents the use of optical alignment devices.

The horizontal carriage, mounted on the support pillar P, can be rotated and braked manually about the axis B–B' (figure 9.4). This rotation is not isocentric but is useful in order to move the couch top away from the isocentre by a distance greater than available by the lateral movement of the couch top. This is particularly useful in the treatment of lateral fields at fixed SSD where the edge of the couch is often close to the isocentre. (For isocentric treatments the centre of the couch is usually close to the isocentre.) The non-isocentric rotation also allows the couch to be swung away from the isocentre to allow access for patients treated on stretchers, hospital beds or wheelchairs, which is sometimes necessary.

### 9.3.4. A novel patient support system

The patient support systems described above are fixed. That is, the moving components are mounted on a frame rigidly attached to the structure of the treatment room and located in a precise position relative to the isocentre. An alternative which has been developed, and which has some advantages, is a robotic patient support system which is free to move over the treatment room floor and whose position relative to the isocentre is continuously measured and controlled by a computer. Figure 9.7 is a photograph of such a system. The couch top is driven up and down between two vertical members in which lead screws provide the lifting mechanism. These lead screws can be relatively light as the only weight they have to lift is that of the couch top and the patient. This is considerably less than the motor lifted weight of the more conventional systems. All other horizontal and rotational movements can be achieved by driving the couch pedestal over the treatment room floor, which therefore has to be extremely smooth and flat in order to maintain the required precision. This is made from a strong epoxy material, which is poured onto the concrete floor slab and, when cured, machined flat *in situ*. There are three pairs of drive wheels on the pedestal, which can be programmed to simulate longitudinal and lateral translations as well as isocentric rotations of the couch. The position and orientation of the couch are measured continuously

134  *Mechanical systems*

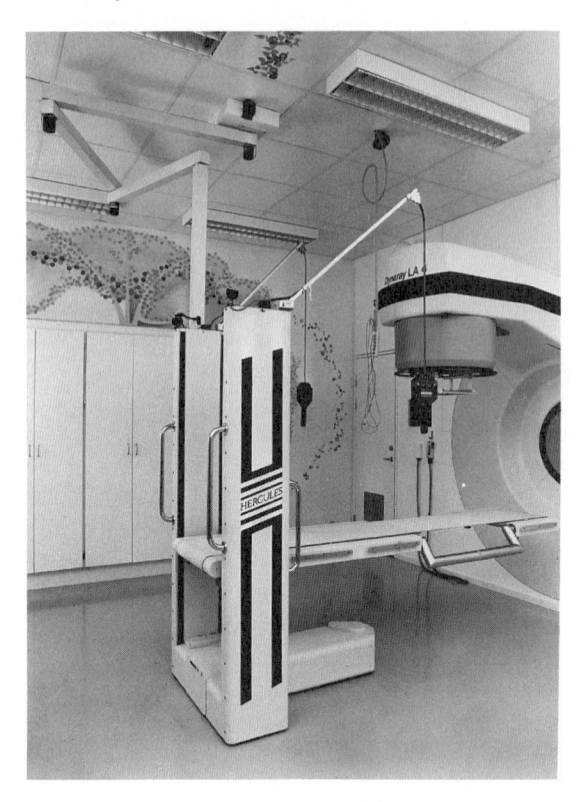

**Figure 9.7.** *A photograph of the 'Hercules', a robotic patient support system. (Courtesy of Precitron AB.)*

from potentiometers actuated by a series of articulating arms which mechanically link the robotic couch to a fixed point on the concrete roof or wall of the treatment room. The control computer calculates the co-ordinates and angles relative to the gantry and the isocentre. It also translates control signals to provide programming of the drive motors to simulate required movements. The system includes software interlocks to prevent collision and can in principle drive the couch round a known obstruction in order to avoid a collision.

Because the system is not rigidly fixed near the isocentre, it can be driven away from the gantry while patients are loaded and can also be driven away if it is necessary to treat a patient on a stretcher or a hospital bed or in a wheelchair.

## 9.4. MOVEMENT CONTROL SYSTEMS

The movement controls for the gantry and couch systems are shown schematically in table 9.1.

**Table 9.1.** *Movement controls.*

| Gantry | Angle | clockwise |
| | | anticlockwise |
| Couch | Height | up |
| | | down |
| | Longitudinal position | +/− |
| | Lateral position | +/− |
| | Isocentric rotation | clockwise |
| | | anticlockwise |
| | Non-isocentric rotation | clockwise |
| | | anticlockwise |

All but the non-isocentric rotation of the couch top are usually servo controlled. Each of the five servo-controlled movements is operated by a continuously variable and reversible input voltage. Hence the system requires five knobs or levers, each controlling a centre tapped potentiometer, or alternatively a smaller number of controls whose function can be switched between the separate movements. The switched arrangement clearly limits simultaneous operation of all the movements. In addition the system requires an 'enable' button, which provides an overall safety control function to all of the five systems. The operation of any movement then requires two actions, operation of the enable button and operation of the selected movement control. This protects the system from casual action by unauthorized persons and from inadvertant operation as a result of a single fault in the control system. A further safeguard, which is easy to engineer in computer controlled systems, is to disable movements at times when they are not required, e.g. during irradiation of a static treatment set-up. Movement controls need to be within easy sight of the patient on the treatment couch and may conveniently be mounted on the couch itself or on a pendant adjacent to the couch. The possibilities of remote control of these variables is further considered in chapter 10 on control systems.

For rotation therapy, the gantry rotation has to be remotely controlled during the radiation. Normally this requires that the dose delivered per unit angle be constant. This can be achieved by stabilizing the dose rate, as discussed in chapter 5, and operating the gantry speed control at a predetermined input voltage, or by making the speed control voltage proportional to the dose rate. Although these have just been stated as alternatives, there is no reason why both should not operate simultaneously, i.e. why the stabilized dose rate signal should not be used to regulate the gantry speed control voltage.

## 9.5. SCALES FOR MECHANICAL SYSTEMS

An angular scale, 0–360°, is required to indicate the position of the gantry. Further angular scales are required to indicate the position of the patient support system, which includes that for the position of the turntable around the isocentre with respect to the gantry axis and the position of couch top with respect to turntable around the pillar P. Linear scales are also required to measure longitudinal and transverse movements of the couch top and vertical movements of the pillar P. The zero positions on the three angular scales should place the gantry with the radiation beam pointing directly downwards, and the long axis of the patient couch placed parallel to the axis of rotation of the gantry.

Although all the distances and angles will be measured and displayed electronically mechanical scales are very useful, particularly during calibration of the electronic systems but also during operation. Although giving an impression of great precision electronically displayed scales are sometimes difficult to use because of quantization effects at the least significant digit and relatively slow refresh rates for the display.

## 9.6. PATIENT SAFETY

The required movements of the gantry and couch are such that it is possible for a patient, or indeed an operator, to be crushed. Such an accident could happen as a result of operator error, or a failure of one of the control systems resulting in one of the motors failing to respond to a stop signal from the operator. There is also the risk of a purely mechanical collision between the couch and gantry resulting in damage to the equipment.

The most common device to deal with these risks is to mount a light-metal 'touch ring' on the end of the treatment head. It can be arranged that this ring holds a microswitch in the closed position and that a deflection of the ring by a few millimetres will return to its normal open position. Opening of the microswitch can disable a contactor to interrupt the power supply to all drive motors. This is a 'fail-safe' system in that a failure of the microswitch to close will de-energize all the motors which could cause a collision. Similar devices using inductive proximity sensing circuits have also been used.

The alternative is to use transducer signals indicating the position of all the couch and gantry motions and feed these to a microprocessor which is programmed to forbid those movement combinations which would allow a patient to be crushed or a collision to occur. This avoids the obstruction which a touch ring system must present to the operator but does not take into account the objects, or indeed the patient, whose size and position on the couch will be variable.

It should be noted that it is difficult, if not impossible, to design a system which is sufficiently sensitive to avoid all collisions without preventing some necessary movements. A compromise between automatic collision avoidance and careful operation by well trained staff is essential.

With either of these systems there remains the possibility of a multisystem failure, resulting in a drive motor continuing to operate in spite of a correct operator action and a subsequent correct operation of the anti-collision system. An emergency system to deal with this remote possibility is required because of the very serious risks involved. This can be provided by bringing all power supplies through a contactor operated by one set of buttons in the treatment room and control area. If all else fails and one of the drive motors is still moving towards a collision, the operator can then de-activate the whole system from one of these buttons.

The risks discussed so far are concerned with faults caused by collision of the motorized systems which are inadvertently powered. There are however a range of risks associated with loss of power. These have to be minimized by careful design. The most catastrophic risk is probably the mechanical failure of a nut driven by the lead screw used to lift the treatment couch. A second nut can be fitted to the lead screw and allowed to rotate with the lead screw as long as the main drive nut is load bearing. If the drive nut fails the 'catch nut' is locked and the worst effects of a failure avoided.

Loss of electrical power or failure of an electrical component can occur when the couch is at such a height as to make it impractical to unload a patient safely. The alternatives are to provide a separate, probably battery operated, motor or a manual drive so that the couch can be lowered to a safe height.

Loss of electrical power to the horizontal movements of the couch should leave the clutch mechanisms engaged. It is then very difficult to unload a patient from a freely floating couch top. The final hazard noted here is that of the potential 'finger trap': a gap between any two parts of the mechanical assembly can close and in doing so crush part of a patient or operator. This is a hazard even if the parts of the machine are moving slowly and under manual operation, as the momentum carried by heavy moving objects such as the couch top (with or without a patient) is considerable.

Much is written about the radiation hazards of radiotherapy but mechanical and electrical hazards probably represent a greater risk, with the mechanical hazards being the most difficult to avoid.

# CHAPTER 10

# CONTROL AND INTERLOCK SYSTEMS

Medical linear accelerators are, as can be seen from the foregoing chapters, extremely complex machines. The control and interlock systems are correspondingly complex and it is neither possible nor appropriate to describe them in great detail as the details are specific to individual manufacturers and models. This chapter will therefore discuss general principles, illustrated where necessary with specific solutions.

Control functions can be divided into those which are basically binary, i.e. on or off, and those which are analogue, i.e. continuously variable. The analogue functions can be further sub-divided into those where a fixed pre-determined level of a controlled variable item is required and those where the variable item is controlled by a feedback signal derived from the controlled function. The control of a particular part of a medical accelerator, such as the operation of the pulse modulator, the electron gun or the positioning of the gantry, may require combinations of all these types of control.

These systems will be discussed under three headings: machine controls and interlocks, safety interlocks and treatment controls and interlocks. Machine controls and interlocks are used to ensure that the accelerating waveguide and its associated components operate according to specification and that it is energized only when it is safe to do so; they are there to protect the machine itself. Safety interlocks are used to protect patients and staff from the many dangers that arise in the operation of these machines. These include mechanical and electrical dangers as well as the obvious dangers familiar in radiological protection. Finally treatment controls and interlocks are used to set up the accelerator to treat each individual patient and to ensure that the radiation output is consistent with the treatment prescription. Particular consideration has to be given to computer control systems as these have facilitated the development of a degree of automation in radiotherapy which although possible with traditional control systems was not widely available. Dynamic therapy in which the geometry of treatment is varied during irradiation is one such application of automation.

## 10.1. MACHINE CONTROLS AND INTERLOCKS

Most of these have been mentioned in discussing the major components and equipment which make up a linear accelerator. The interlocks are summarized in table 10.1, which splits them into 'start-up interlocks', those with transducers which indicate that the parameters listed under this heading are within acceptable limits before it is possible to energize the modulator to supply microwave power to the accelerating waveguide, and 'operation interlocks', which, in addition to those above, continuously monitor the systems during irradiation. Clearly the start-up interlocks continue to operate during irradiation so if, for example, the vacuum pressure rises during irradiation the machine will automatically stop and all components whose operation is dependent on the vacuum interlock will be turned off.

Some of these interlocks have not been mentioned previously. The focus and bending current interlocks prevent acceleration of the electron beam without focusing and bending magnetic fields as this would cause the beam to strike the accelerator or bending chamber walls. The result would be risk to the vacuum system and the possibility that high levels of leakage radiation would be produced. The 'warm-up timer' is required because the cathodes of the high-powered valves (the thyratron and the magnetron or klystron) require a heating period of 5–10 min before they come up to operating temperature on initial switch-on.

Interlocks are provided to ensure that the accelerator, or any part of it, will operate only if the correct conditions are satisfied by preparation (e.g. by prior evacuation of the waveguide), by correct selection of machine settings (e.g. by choosing the correct beam current for electron therapy) and by correct running of the machine (e.g. by monitoring the beam flatness). A well designed machine will take account of all possible variations of operating conditions including those that can be incorrectly selected by the operator and many predictable faults. It is a reasonable expectation that such an interlock system will prevent serious injuries to patients and staff and major damage to the machine. However there will always remain the possibility of component failure, the effects of which are minimized by interruption of the power supply by the use of fuses or current sensitive circuit breakers. In most cases the loss of power resulting from a blown fuse will cause an interlock to operate but fuses should not be regarded as interlocks *per se*.

Mechanical limit switches are provided to prevent travel beyond the range of motorized movements. The primary purpose of these is to prevent damage to the equipment. These interlocks must be direction sensitive so that, having reached a limit, the motor can be reversed.

## 10.2. SAFETY INTERLOCKS

Electrical safety interlocks are required to protect staff from high voltages, and are normally operated by switches which open when the doors of the relevant cubicles or cabinets are opened. Operation of these switches will then prevent the appropriate mains contactor from being energized or drop it out if it is already energized.

Mechanical hazards have been mentioned in chapter 9. Mechanical safety controls include collision detectors operated when the couch or patient comes into contact with the treatment head. It should be noted that when accessories, such as electron applicators, are attached to the head it is necessary to extend the collision detection to the accessory, which, in the case of an applicator, will be brought close to contact with the patient.

The room interlocks are required to avoid the possibility of operation of the machine with personnel, other than the patient, inside the treatment room. An interlocked barrier is installed at the entrance to the treatment room maze. It is broken when someone enters the treatment room and must be manually reset before irradiation can take place. The interlock may be in the form of a door operated switch, or a light beam projected across the room entrance and detected by a photocell, which in turn operates the interlock. These devices are simple in principle, but in practice the requirements for such safety devices are very stringent. To prevent a single failure compromising the safety function two such switches are used to provide redundancy. If electromechanical switches are used they must be of a type that have positive actuating mechanisms where the contacts move directly with the actuating lever rather than indirectly as is the case with most microswitches. Optical interlocks also require redundancy and must be fail safe, continuously monitored and not affected by ambient light.

Each time the room interlocks are operated a positive action, e.g. shutting the door or resetting the photocell interlock, is required to reset them. Ideally resetting the room interlock should incorporate a 'search and lock-up' procedure. This forces the operator to enter the treatment room and check that it is empty before operating a control such as a push button switch inside the room: this is the first, 'search', stage of the process. The second, 'lock-up', stage is to complete the reset sequence by operating a second push button switch outside the room within a pre-set time of typically 15 s. A further refinement is that an alarm sounds inside the treatment room after the first stage so that anyone not seem by the operator is warned of the intended irradiation. The person setting the room interlock is then accepting responsibility for the decision that no one, other than a correctly set-up patient, is in the treatment room before the machine can be energized. Manual operation of this reset is one of the few positive actions required from the machine operator in relation to the machine and safety interlocks listed in table 10. l. Most of the interlocks are monitored continuously and are automatically set if safe operating conditions are in place. It is necessary for the status of all the interlocks to be indicated so that in the case of a fault the equipment can be restored to its correct state.

**Table 10.1.** *Interlocks.*

|  | Start-up interlocks | Operation interlocks |
|---|---|---|
| Machine | Vacuum pressure<br>Gas pressure in pressurized waveguide<br>Cooling water flow<br>Cooling water temperature<br>Focus coil currents<br>Bending magnet current<br>Warm-up timer<br>Mechanical limit switches | Beam flatness<br>Beam energy<br>Excess dose rate<br>Modulator current overloads |
| Safety | Electrical safety interlocks<br>Room interlocks<br>Collision detectors |  |
| Treatment | Treatment settings<br>  radiation type/energy<br>  static/dynamic wedge<br>  applicator/accessory<br>Dosimetry settings<br>  monitor units<br>  time<br>Movements<br>  movements ready |  |

## 10.3. TREATMENT CONTROLS AND INTERLOCKS

These relate the machine settings to the needs of an individual patient via the radiotherapist's prescription and are discussed here for a multipurpose machine operating either for x-ray therapy at two photon energies or for electron therapy over a range of electron energies. The interlocks required for a simple machine used exclusively for x-ray therapy at a single energy are clearly a subset of those required for the more versatile machine.

### 10.3.1. Treatment prescription

As the purpose of treatment interlocks is to ensure that the machine operating conditions are consistent with the treatment prescription for each individual patient it is necessary to consider the contents of such a prescription.

Table 10.2 shows the various items required for patient prescription together with the machine controls and interlocks which have to be set in order for the machine to deliver the treatment.

(i) *Radiation type.* The choice of radiation type is a binary decision but involves the operation of many controls and interlocks. The target, flattening filter and scattering foils are usually driven to the appropriate position, in or out of the beam, by motors or other electromechanical

**Table 10.2.** *The treatment prescription and its relationship to machine controls.*

| Prescription item | Machine setting | Machine control function |
|---|---|---|
| **1** *Patient identification* | not generally applicable | input for automatic set-up used for verification etc |
| **2** *Radiation type and energy* | | |
| X-rays | target in | interlock |
| | correct flattening filter | interlock |
| | selected beam energy | pre-set and servo controls of beam production |
| Electrons | target out | interlock |
| | correct scattering foils | interlock |
| | selected beam energy | pre-set and servo controls of beam production |
| | applicator | interlock |
| **3** *Irradiation technique* | | |
| Static | select static treatment | interlock |
| Rotation | selected rotation speed and direction | interlock and proportional control of gantry speed |
| **4** *Wedge filter* | | |
| Fixed wedge | wedge number | interlock |
| Universal wedge | dose with wedge dose without wedge | interlock and control of two dose segments |
| Dynamic wedge | collimator trajectory | interlock and proportional control of collimator positions |
| **5** *Field size* | | |
| X-rays | collimator setting | manual or automatic setting |
| Electrons | correct applicator | interlock and automatic setting of x-ray collimator |
| **6** *Field shape* | | |
| X-rays | correct shielding block | interlocked if coded |
| Electrons | correct applicator end frame | interlocked if coded |
| **7** *Field orientation* | collimator angle | manual or automatic setting |
| **8** *Field locations* | not usually defined | manual setting of isocentric gantry and couch |
| **9** *SSD* | not usually defined | manual setting of isocentric gantry and couch |
| **10** *Dose* | monitor units required for each irradiation segment | interlocks for dose difference, excess dose rate termination on correct dose delivered |

actuators and their position monitored and interlocked. It is important that the control and interlocking functions are separated. For example if a microswitch is used to detect the position of a scattering foil and to stop the drive motor when the foil reaches the correct position a second switch is required to provide the interlock function, checking that the foil is in the correct place for the electron energy selected. The use of a single switch would lead to a situation where a single fault would compromise safety.

The selection of radiation energy determines the choice of flattening filter or scattering foil in the treatment head and is used to preset the operating conditions of the accelerator and the beam transport system so that a beam of the correct energy is produced. If the dose control systems have switched calibration settings for each energy these are also controlled by the selection of energy.

(ii) *Irradiation technique.* On machines capable of rotation therapy the irradiation technique must be selected and in the case of rotation, the simplest form of dynamic therapy, it is also necessary to select the rotation speed and direction together with the initial and final gantry angles. Rotation speed can be specified either in terms of degrees per monitor unit or degrees per minute in order to deliver the required dose over a defined angle. Specification in degrees per monitor unit ensures that the required dose is delivered over each part of the arc of rotation independent of the dose rate, which may not be constant during exposure. Alternatively, if the dose rate can be precisely controlled the gantry can be rotated at a specified number of degrees per minute and the dose rate varied to compensate for any variations in actual speed in order to achieve the same result. As the machine parameters set for rotation therapy are not independent it is necessary to interlock against incompatible settings: for example, if the start and stop angles required an anti-clockwise rotation then selection of a clockwise rotation should inhibit irradiation and provide an indication of incompatible settings.

(iii) *Wedge filters.* Clearly the control functions required for wedge filters are dependent on the way in which wedged beams are produced. These have been discussed in chapter 6.4. Physical wedges have to be interlocked to ensure that the wedge is in place when, and only when, required by the prescription and also to ensure that the correct wedge angle is used. In the case of removable wedges this can be achieved by coded switches of any type including microswitches, opto-electronic encoders and magnetically operated reed switches. The wedge angle using a universal wedge is determined from the proportion of dose delivered with the wedge in place. Therefore in this case the control information must be common to the dosimetry system to ensure that the wedge is removed after the predetermined dose has been delivered. A further desirable interlock for wedged beams is to be able to interlock against the prescribed collimator–head angle as this determines the orientation of the wedged field relative to the patient. Finally, as most physical wedges cover a field size smaller than the maximum available, it is necessary to arrange an interlock to prevent irradiation when the maximum wedge field size is exceeded.

(iv) *X-ray field size.* X-ray field size at the patient is determined by collimator settings which take into account the SSD. As has been stated in chapter 8 the collimator settings are usually scaled to be equivalent to the field size at the source to isocentre distance. Most collimators are motor driven, opening up the possibility of automatic setting, and are controlled by switches, in which case there has to be a compromise in choosing the speed of control. It must be fast enough to effect a large change in field size in a reasonable time, say 20 s, but slow enough to enable a reasonably well co-ordinated operator to set the prescribed field size to within 1 mm. Variable speed control of the collimators, using a potentiometer as the control element, avoids such a compromise at the expense of some additional complexity.

Electron field size is determined by the physical dimensions of the electron applicator. The presence of an applicator on the treatment head generates an interlock signal so that all the machine settings for electron therapy have to be selected in order to permit irradiation. In addition the electron applicator can provide a control signal to the x-ray collimators so that they can be automatically set in the optimum position for each applicator.

(v) *Field shaping.* In general field shaping devices, which include compensators, are attached to the treatment head manually. However as these devices can have a significant effect on the radiation output, it is desirable that they should also be interlocked.

(vi) *Field orientation, location and SSD.* Most of the prescription items give rise to machine settings needed to control the machine or to interlock its operation. However the location of the patient in the frame of reference of the treatment machine and the location of the beam relative to anatomical landmarks on the patient are not easily related to machine settings. In some cases this is overcome by immobilization and fixation of the patient to the treatment couch, in which case the SSD and field location can be uniquely related to machine settings. However, in general, the correct position of the patient is dependent on the professional skills of the radiographers, using manual controls to position and orientate the couch gantry and collimators.

Perhaps the most fundamental parameter in the treatment prescription is the dose which the patient is to receive. The control of dose has been discussed at some length in chapter 7.

## 10.3.2. Operation of treatment controls

So far the requirements for interlocks and controls, in relation to the operation of the machine and the delivery of the treatment prescription, have been described. The provision of these controls can be further illustrated by describing, in order of increasing sophistication, the ways in which machines can be operated.

*10.3.2.1. Manual operation.* The machine operator sets up the patient on the treatment machine in accordance with the machine settings that have been determined from the prescription. Assuming that the machine interlocks are all in the correct state, the system is ready for irradiation which will start when the 'on' button is pressed. The linear accelerator is then under the control of the dosimetry system, as previously described in chapter 7, but this can be overridden by a manual control which allows the operator to terminate or interrupt the exposure if, for example, the patient is seen to move. Normally the dosimetry system will terminate the exposure when the pre-set number of monitor units have been delivered.

*10.3.2.2. Select and confirm operation.* In the discussion of treatment control and interlocks, above, it was tacitly stated that various settings should be interlocked without explicitly stating that they should be interlocked against the prescribed settings and without acknowledging that the prescribed settings have to be loaded into the machine in a way that is independent of the actual settings used for irradiation. The second level of sophistication, after simple manual operation, requires two steps in setting a control item.

In these systems the patient is set up for treatment as before and the transducers which pick up the machine parameters selected, for example the wedge angle chosen. The operator, having set up the patient for treatment, then has to enter the corresponding values for the machine parameters by hand at the control console. The system will only allow the irradiation to be started if the entered parameters and the measured parameters are in agreement.

Field shaping devices, item 6, table 10.2, can be included in an electrical select and confirm system by mounting them on Perspex plates, which attach to the treatment head and can be coded by the use of a suitable transducer. Alternatively, the use of such devices can be regarded as selected and confirmed by the use of the main optical beam. If a specially cast beam shaping block or a standard block or arrangement of blocks are used and a shadow outline marked on the patient's skin, then only the correct block arrangement can be confirmed by the optical beam.

The use of measured gantry and couch parameters, items 8 and 9, table 10.2, to confirm the beam direction on a patient will only be useful if the patient is placed in a defined position on the treatment couch. Since this position cannot be defined with complete precision, some tolerances have to be built into the select and confirm system to allow for this.

This system can be applied to all or any combination of the machine parameters. Select and confirm systems for the use of wedge filters have been in use for many years because the use of the wrong wedge can introduce very serious dosimetric errors. While all the machine parameters could be included in a manual select and confirm system, operation of the equipment would be extremely laborious and time consuming. In practice, manual select and confirm is usually limited to radiation type, wedge filter systems, the dose monitoring system and the start and stop angles for arc therapy treatments.

This might be described as the traditional method of controlling a linear accelerator and correct usage of the equipment then depends entirely on the radiographer's skill and experience.

*10.3.2.3. Automatic select and confirm—verification.* If the information on the treatment sheet includes the machine parameters and these are also available in a form which can be read directly into the treatment unit, for example, in a data file on a floppy disk, then the confirm element of the select and confirm procedure can be automatic. In other words, when the patient is set up for treatment, the irradiation can only start when the machine parameters agree, within defined tolerances, with the values in the data file which is itself linked to the treatment record. Entry of the machine parameters to the data file requires an initial set-up of the patient so that their values may be determined. This can be performed on the machine itself, or on a simulator (see chapter 15). The patient's identification will be stored in the data file and the logic of the system would also require that the use of the correct treatment sheet be confirmed, maybe by identification by a bar code. This can be done by entering the patient's identification in the control system and then requiring a coincidence between this information and that on the bar code. The verification that the correct machine settings have been used for each individual patient is one aspect of the wider topic of verification which is covered further in the next chapter.

*10.3.2.4. Automatic and assisted set-up systems.* If it is accepted that the necessary information for a treatment can be stored as in the previous paragraphs, then it is feasible to arrange for the machine to set itself up automatically. The procedure is then for the radiographer to place the patient in the correct position on the treatment couch, enter the treatment sheet into the control unit, and confirm the patient's identity. The machine can then automatically set up the machine parameters listed in table 10.2 by what is still a select and confirm procedure, in which signals from the machine transducers are compared with the stored information. This system can include the setting up of a wedged field if a motorized universal wedge is used but obviously not if manually positioned wedges are used.

It is useful to distinguish automatic set-up from assisted set-up. Some items are fundamentally precise and can be set up automatically. These include the radiation type, the radiation energy and the number of monitor units required. The value required is known from the treatment data file with absolute precision. Other items, which tend to be those with continuously variable parameters, are less well defined. These can be set to the prescribed value, which is only approximate and might require some fine tuning for each irradiation. These include most of the mechanical settings such as the couch positions, which are variable because the patient will not be in exactly the same position for each irradiation, and the field size, which might need fine tuning to align the field edges with skin marks. One definition of assisted set-up is therefore semi-automatic set-up, in that it requires manual fine tuning of certain items. A second definition is that it is automatic set-up requiring operator supervision and intervention. For example, automatic movement of the couch and gantry are potentially dangerous operations. An assisted set-up function whereby the couch and gantry are automatically moved to the prescribed positions under the control of a single 'enable' button provides the advantages of speed and accuracy while avoiding some of the dangers arising from operator limitations.

### 10.3.3. General comment on machine operation

The three systems described for controlling the delivery of a prescribed treatment to a patient all require the preparation and checking of a treatment sheet. The responsibilities and skills required to do this are just as great for the automatic systems as for the manual systems.

## 10.4. CONTROL AND INTERLOCK CIRCUITS

This chapter has so far discussed the principles involved in operational use of control and interlock systems. It is now necessary to consider how these systems relate to the circuits discussed in the previous chapters. Different types of control circuit are considered in each of the following sub-sections, where a specific example is used for illustration.

### 10.4.1. Interlock circuits for radiation control

After all the necessary actions to comply with the instructions on the treatment sheet have been carried out, the radiation is switched on by energizing the modulator, which has been described in chapter 3.5. This can be done by switching on the high-voltage supply for the pulse forming

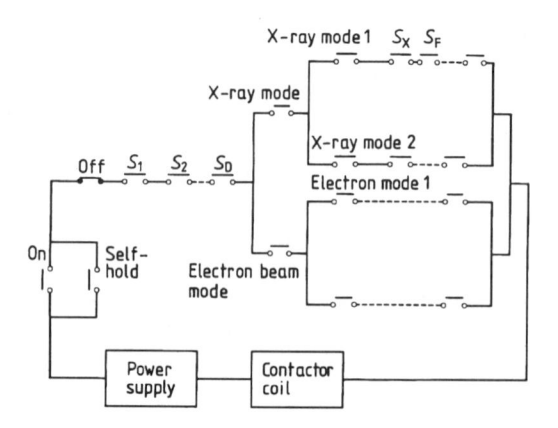

**Figure 10.1.** *The control circuit to energize the contactor coil, which starts irradiation.*

system i.e. by closing the main three-phase contactor for the high-voltage supply circuit and the applying pulses at the required PRF to the thyratron.

In principle the circuit to switch on the modulator is shown in figure 10.1. Switches $S_1$ and $S_2$ are two of what might be many contacts, which are closed if the interlocks in table 10.1 indicate that the items listed are operating or have been selected correctly. They may be switches, which monitor particular items directly, or relay contacts, which are indirect but may be used to monitor a group of individual items, in which case the system can be considered to be a hierarchical structure. The primary, or top-level, chain of contacts can also include a key operated switch, providing a degree of security as the system can be made inoperable by removal of the key. When the dose monitor system is correctly set up it will indicate 'dose monitor ready' and close contacts $S_D$. Where an alternative has to be chosen, e.g. x-ray mode or electron beam mode the contact chain branches, and if the system allows the choice of more than one x-ray quality the circuit branches again, for example, into x-ray mode 1 and x-ray mode 2. Each of these sub-branches consists of a series of contacts which have to be closed as a consequence of the choices made. For example, if the choices made are x-ray mode, then x-ray mode 1, the contacts $S_x$ and $S_F$ are closed when the x-ray target and the appropriate beam flattening filters are moved into position. The circuit can have any number of branches and the branches can have any number of sub-branches, where each select and confirm action requires two contacts in series to close. As interlock chains become very long, that is that they include a large number of individual interlocks, it is convenient to split them up into groups, each group of interlock switches in series causing a relay to close and the relay contact then providing a single contact in a shorter primary interlock chain. Correct operation of the system will close all contacts $S_1, \ldots, S_D$ and all those in the branches and sub-branches chosen. The contacts on the 'off' button are normally closed, so operation of the 'on' button will energize the modulator and start an irradiation. The 'on' button has parallel 'self-hold' contacts, which can be low-power ancillary contacts operated by the HT contactor coil. This ensures that the irradiation continues when the button is released. The radiation may be switched off by opening the 'off' contact, or by the opening of a contact by the dose monitor indicating 'irradiation complete', or by the opening of any of the contacts indicating a fault condition including unexpected operation of the room door interlocks. If the exposure is terminated for any of these reasons, the self-hold contact in parallel with the 'on' button will also open, so that exposure does not restart automatically when the contact is reset.

Emergency off switches, which are accessible in the treatment room, in the control area and any other spaces where equipment is housed, must be able to stop irradiation. They are not connected directly to the radiation control interlock chain as described here because radiation is just one of the hazards to be protected against by the operation. The emergency switches are therefore configured to remove power from all circuits, not just those needed for radiation control.

The circuit just described ensures that the logical sequence of tests and decisions that are required have been carried out before an irradiation may be started. These functions may be carried out by a chain of electromechanical contacts as described, or by the use of solid state logic circuits, or by software logic in a microprocessor controlled machine.

Interlock chains for multipurpose machines can become extremely complex with many hierarchical levels and multiple branches. Interlock design is critical to safety and every possible combination of operational states must be examined to ensure that faults, or detection of unacceptable conditions, in one part of the circuit do not have ramifications in another. The interlocks should be designed to fail safe both individually and as a complete system.

All interlocks must be operative while the machine is being used in the clinical mode, i.e. for the treatment of patients. However the requirements can be less stringent in the service mode: this is necessary in many cases to allow for service procedures and adjustments to be carried out. Selection of the service mode of operation therefore disables or bypasses some of the interlocks but not, of course, those concerned with safety.

In current practices most machines use hybrid circuits, a mixture of electro-mechanical contacts, solid state logic and software logic. It is, however, considered to be good practice to maintain hard-wired interlocks for the main safety critical systems such as those allowing irradiation to be initiated and terminated. Software systems will often be used to monitor the operation of such circuits and might also be capable of terminating irradiation but the most direct method of physically opening a pair of contacts is the one which provides the highest level of confidence and for which risk analysis is the most straightforward.

### 10.4.2. Analogue control circuits

Many different types of circuit are needed to provide analogue control for the sub-systems which make up the treatment machine. In this context 'analogue' is used to describe the continuously variable nature of the output of the circuit rather than the technology used to achieve it. For example, the mean current supplied to a motor might be controlled by pulse width modulation using digital electronic components (gates, monostables etc) but the mean current could be continuously variable and therefore an analogue output.

*10.4.2.1. Stabilized power supplies.* The simplest analogue controls are those required to deliver current to components such as the focus coils or the magnetron filament. Although these currents might need to be set at different levels for each set of operating conditions the levels are in principle constant. The control is achieved by selecting pre-set voltage levels from a variable power supply, maybe by switching one from a bank of variable resistors or potentiometers into the control input of the power supply. If necessary the current can be monitored and the feedback signal generated compared with the (pre-set) reference level in order to provide further stabilization.

In the first example given the pre-set focus currents will depend on the electron energy and will therefore be switched as part of the process of selecting the energy. In the second, the magnetron filament current will depend on the power level at which the magnetron is operating and the preset control signal will be derived from other parts of the accelerator such as the pulse modulator (figure 3.10).

*10.4.2.2. Variable supplies.* Motors are used in many parts of the linear accelerator and include a range from small motors, whose speed is not critical, to large motors, where both the required speed and load is variable. If speed is not critical or the load is constant a motor can be controlled with a fixed or variable voltage supply. However in the case of motors subjected to variable loads, such as the couch lift motor, it is necessary to compensate for the back-EMF generated by the motor changing with the load. Such a circuit, illustrated in figure 10.2, utilizes a combination of voltage and armature current feedback to provide a correction to the main control voltage derived from a potentiometer. Note that if the mean power is controlled by pulse width modulation (PWM) two advantages over linear, or voltage, modulation are obtained. Firstly the torque generated by the motor is constant and therefore fine control at low speed is more easily achieved. Secondly the power transistors in the output stage of the motor control unit

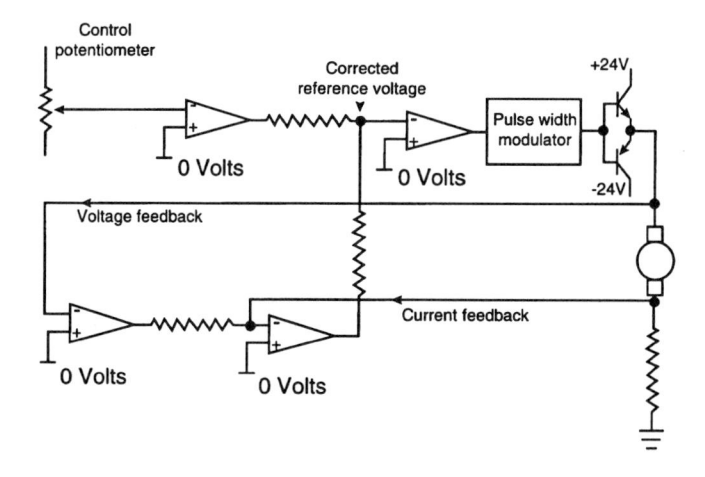

**Figure 10.2.** *A motor control circuit with armature voltage and current feedback.*

are either turned fully on or fully off and therefore power dissipation in these components is minimized.

*10.4.2.3. Servo controls.* Many of the functional control items in a linear accelerator are controlled by electrical signals. These include mechanical items such as the position and speed of the gantry, electron beam items such as the position of the electron beam as it strikes the target and microwave power items such as the operating frequency of the magnetron. This short list of examples, set out in table 10.3, is not exhaustive and many more can be found in the previous chapters. The operating conditions of these items can be measured by appropriate transducers, which are given in the table: the signal generated can be compared with a reference value and a control voltage or current produced to minimize the difference. When the difference is zero the item has been controlled to its required state. Control systems including feedback links, which may include mechanical elements, are called servomechanisms, or servos from the Latin word 'slave'. This is because the output of the control transducer, the slave, is forced to follow the input from the reference signal, the master.

The servos for microwave frequency and electron beam position have been described in chapters 3.4 and 5.3. A servo for gantry speed, which is necessary for rotation therapy, is illustrated in figure 10.3. The required speed (degrees per minute) is determined for each treatment by two factors, the number of degrees per monitor unit, calculated so that the prescribed number of monitor units are delivered over the prescribed arc of rotation, and the dose rate (monitor units per minute), which is variable and can be measured from the beam monitor.

152  *Control and interlock systems*

**Table 10.3.**  *Examples of servo controls.*

| Control item | Source of control signal | Control mechanism |
|---|---|---|
| Gantry angle | potentiometer geared to gantry | gantry drive motor |
| Gantry speed | tacho generator geared to gantry drive motor | gantry drive motor |
| Electron beam position | sectored ionization chamber | beam steering coils |
| Microwave frequency | diode probes in tuned cavities | Magnetron tuner position (motor) |

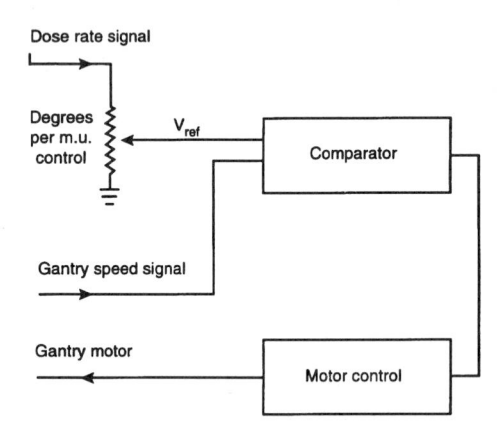

**Figure 10.3.**  *A servo to control the gantry speed during rotation therapy.*

As the dose rate varies it is necessary to control the gantry speed so that it varies in a proportional way. The reference signal for rotation therapy is derived from a potentiometer (in practice this can be a network of fixed resistors with switching to simulate a potentiometer) fed by the dose rate output signal from the beam monitor circuit. The potentiometer is calibrated so that the wiper voltage $V_{ref}$ is proportional to both dose rate and speed ($°$ $MU^{-1}$) This is compared with the control signal from a tacho generator driven from the gantry drive motor and which is proportional to the gantry speed. The difference, or error signal, is used to increase or decrease the output of the motor control power module and so completes the servo feedback loop.

## 10.5. COMPUTER CONTROL

Computer control of linear accelerators has developed rapidly to a point where most machines supplied world-wide include some computer control. However the extent to which computers are used covers a wide range, from those where computers, used to store and monitor data, are interfaced to accelerators, to those where integration is almost complete and all aspects of their machine control systems are handled by computers. Although the distinction is somewhat arbitrary the following sections deal with machines interfaced to computers and machines controlled by computers.

### 10.5.1. Machines interfaced to computers

The most common early applications of computers to the delivery of treatment, as opposed to the planning of treatment, was for verification and recording. In these systems a small computer was interfaced to the accelerator so that the operating parameters, particularly those relating to the treatment prescription, could be compared with stored values of the same parameters. The control function was exercised by an interlock (a relay contact) being closed by the computer if all the measured parameters were correct, within a predefined tolerance. The number of parameters monitored was initially limited by the number of data that could be stored on the media available. This was a serious constraint when the data capacity of a magnetic strip was 256 bytes, which can be compared with more than 1 Megabyte on a standard 3.5 in floppy disk. Clearly current systems are not limited by data storage and the control functions have become more sophisticated. At the treatment control level the monitoring function has been complemented by automatic and assisted set-up of treatment parameters as has been described above.

At the machine control level there are several ways in which an interface computer can enhance the performance of a linear accelerator's control system.

Interlocks can be monitored, and non-critical interlocks operated, via the computer. Additional functions can be included within the overall interlock scheme without necessarily increasing the complexity of the 'hard-wired' interlock chains. It is often necessary to have different interlocks operating at different times, e.g. for clinical operation and for testing. In this case elements of the interlock system that are carried out within software can effectively be reconfigured by invoking different programs.

On multipurpose machines it is necessary to switch control signals (e.g. preset input voltages to stabilized power supplies and servos etc) according to the mode of operation selected. If this is carried out by computer, problems associated with the operation of large numbers of electromechanical contacts can be avoided. For example, in a machine with eight modes of operation (two x-ray energies and six electron energies) the reference voltage for one function, such as the beam bending current, needs eight potentiometers and an eight-position selector switch or eight relay contacts in a mechanically switched system. Alternatively the reference voltage can be generated from a digital to analogue (D to A) converter and 'switched' by loading the appropriate digital input signal to the D to A converter from a database. The adjustment of the reference voltages is carried out by editing the data base rather than adjusting the potentiometers with the added advantage that settings can returned to their original values with confidence.

The options for presentation of information are also much enhanced by the use of computers. There is a vast amount of information available but it is not all required all the time. For example the information needed during set-up, i.e. while the patient is being positioned and the dose control parameters loaded, is somewhat different from that required during irradiation. The use of computers makes it very easy to switch between different displays. Most systems allow individual users (hospitals not individual operators) some freedom to customize the displays to meet the needs of their particular practices. Customizing to accommodate terminology and language in use in each country is an important requirement.

### 10.5.2. Machines controlled by computers

Full computer control of a linear accelerator provides many possibilities not easily available with conventional control systems. These include

(i) the ability to perform tasks that would be impractical with manual control because of the number of data that have to be processed e.g. the implementation of complex dynamic therapy,
(ii) the generation of much more specific reports of machine status, e.g. display of messages concerning abnormal or fault conditions, aimed at rapid fault diagnosis and repair,
(iii) the adjustment of operating conditions by editing a data block rather than, for example, large numbers of potentiometers (such adjustments can be precise, can be automatically logged and can be returned to exactly their original settings),
(iv) the implementation of 'intelligent' interlocks and other control functions, which can change their characteristics depending on the state of the machine (this includes more rigorous cross checking of machine status than is possible with hard-wired circuits),
(v) implementation of automatic and assisted set-up, verification of treatment settings and recording of treatment delivery and
(vi) modification of the linear accelerator's functional capabilities through software rather than hardware changes.

Although the purpose of computer control is to improve the performance and efficiency of the accelerator the methods of achieving it are rather complicated for many reasons, including the following. The linear accelerator, particularly the pulse modulator, is a hostile electronic environment, which must be effectively isolated from the computer hardware. Software for real time safety critical control applications is very

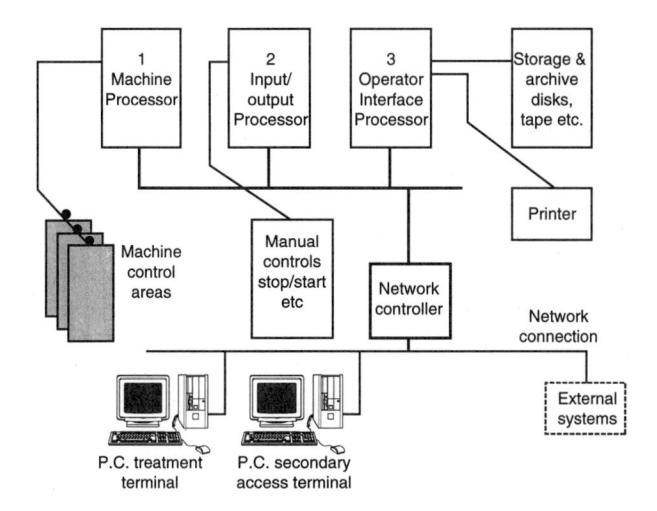

**Figure 10.4.** *An accelerator control system based on multiple processors. (Based in information provided by Elekta Oncology Systems.)*

difficult to produce and even harder to validate. As the performance of the machine is critically dependent on the stored data it is necessary to provide absolute security against corruption of that data.

Some aspects of computer control can be illustrated by reference to the Philips SL series accelerators which were first introduced in 1985. This machine is controlled by three separate processors (figure 10.4), each dedicated to particular control functions but able to communicate with each other and to access common memory so that data generated by one of the processors can be used by another.

Processor 1 provides the interface to the linear accelerator: it sends information and instructions to the three 'control areas' of the accelerator. Each 'control area' consists of a rack of printed circuit boards providing the interface between the computer and the devices that it controls. One control area covers the radiation head, a second covers HT circuits and a third covers all other parts of the machine.

Processor 2 converts the prescription values, from processor 3, into accelerator control values and provides the necessary control functions.

Processor 3 provides the operator interface and deals with the database containing the individual patient prescriptions. Communication is via a terminal consisting of a VDU and a keyboard. It also has interfaces to a printer and other peripheral devices including those for back-up and archiving.

Although the processors provide the control functions in software the input and output originate from the control areas. These include multiplexers, A to D converters (twelve bit) for input, D to A converters (eight and twelve

## 156  Control and interlock systems

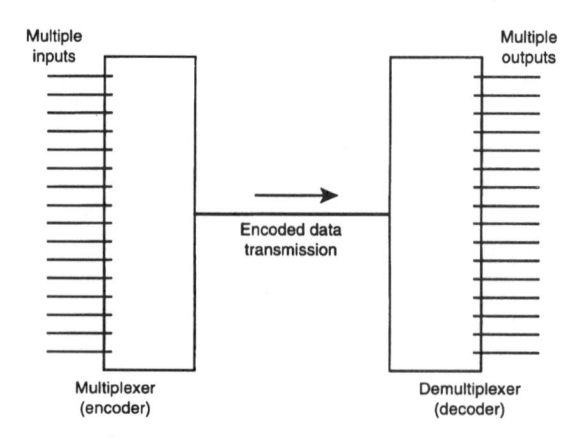

**Figure 10.5.** *The use of multiplexed data for transmission.*

bit) for output, digital input encoders and relay output cards whose precise functions are programmed by instructions passed from processor 1.

Although it is beyond the scope of this book to describe the operation of these devices in detail the purpose of each will be described briefly. More details can be found in general textbooks on computer control engineering and specifically in the training and operation manuals provided by the manufacturers of linear accelerators.

A multiplexer is a device which allows multiple signals to be encoded and transmitted along a single wire. The complementary device is a demultiplexer, which decodes the signals so that they can be distributed to the parts of the circuit where they are used. Figure 10.5 illustrates how the use of multiplexers for transmission of data between parts of a complex system greatly reduces the number of interconnection cables and connections required. The most common 'day to day' application of this technique is in telephone communication, where large numbers of signals are transmitted, apparently simultaneously, over single connections.

A to D converters convert analogue signals, e.g. a current or voltage, into a digital form. The resolution of an A to D converter depends on the number of 'bits' in the digital output. An eight-bit A to D converter, illustrated in figure 10.6, produces an output in the range 0–255 so the resolution is one part in 255. Greater resolution can be obtained by use of a twelve-bit A to D converter, which has a resolution of one part in 4095. In addition to the resolution, determined by the number of bits, it is necessary to consider the scaling between the input and output, e.g. if a signal of 10 V is converted to give an output of 4000 from a twelve-bit A to D converter the voltage resolution is 2.5 mV. The best resolution is obtained by scaling the output to be near to full range for the highest signal that will be handled. A to D converters are used to interface signals from

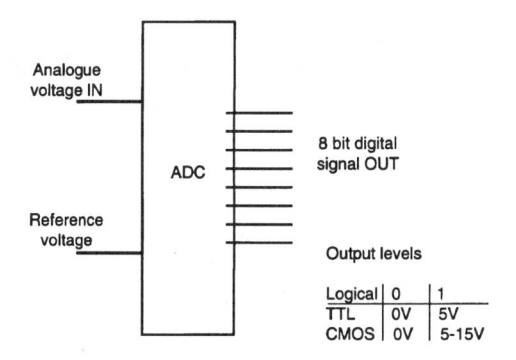

**Figure 10.6.** *A to D conversion.*

the transducers used to measure control variables to the computer control system. A simple example of a control variable is the gantry angle, in which case the transducer could be a potentiometer connected to the gantry drive mechanism. The complementary device to an A to D converter is a D to A converter, which, as its name implies, converts a digital input signal to an analogue output. D to A converters are used to interface digital signals from the computer control system to various component parts of the accelerator, e.g. for the computer control of a variable power supply.

Digital input encoders are interfaces to allow for signals from switches and relays to be monitored by the computer. It is convenient to encode the state of many switches into a single digital 'word', which can be decoded within the computer system. Many of the digital inputs are monitored by the computer and the appropriate action taken after the data has been processed by the control system. In this case the digital input encoder is a passive device. However some inputs require immediate attention, for example to disable an interlock if a particular switch is not operated correctly. In this case the digital input encoder includes logic circuits which carry out the data processing locally as well as providing the encoded data for monitoring by the computer.

Relay output cards allow the computer to control the state of relays, which are used to set various interlocks and control the operation of parts of the accelerator. A simple example could be the 'on/off' control of the beam defining lamp.

In addition to these circuits, which are required in each of the control areas, there are some circuits that have more specific functions including the dosimetry circuits and servo input circuits. As has been noted in section 10.5.1 in relation to safety critical interlocks the dosimetry circuits have been retained as electronic circuits and not replaced by software processes although they are monitored by the computer and some of the functions, such as the dose calibration factors, are controlled by the computer.

Communication between the processor and each control area is by two serial links. The first sends data to the control area and the second sends back an echo and a signal to acknowledge successful communication. These links are momentarily blanked during the magnetron pulse to prevent RF interference.

Logically the accelerator can be considered to consist of a very large number of items (>500). An item may be the state of a switch or relay, the voltage at a particular point in a circuit, a signal from a transducer etc. These items are 'real', in that in principle they could be monitored externally by use of test instruments. Other items are 'virtual', in that they exist only as data items in the operation of the control software, for example the difference between two real items could be computed in order to emulate servo control but this item would not be converted to a measurable voltage or current in the hardware.

The overall control process deals with each item as necessary according to a predetermined priority. The most critical items are processed synchronously, every control cycle, whereas less critical items are processed less frequently and in some cases asynchronously, i.e. when the processor has time. A control cycle consists of three stages. Data are read, processed and then transmitted. There are therefore three corresponding types of item: receive items, which correspond to inputs to the control system from parts of the accelerator, process items, which are items generated from software routines, and transmit items, which correspond to outputs from the control system.

Figure 10.7 shows how one of the beam steering servos can be implemented by computer control. It can be compared with the conventional

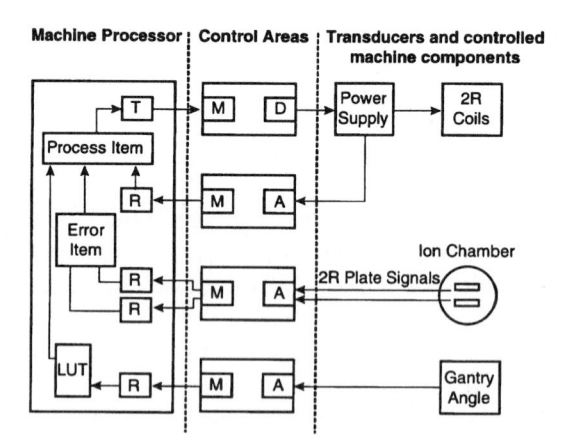

**Figure 10.7.** *Computer implementation of the beam steering servo. A, A to D converter; D, D to A converter; M, Multiplexer; R, 'read item'; T, 'transmit item'; LUT, look-up table.*

servo described in chapter 7.2.1 and illustrated in figure 7.2. The required steering current depends on four receive items: the gantry angle, two beam uniformity signals from the ionization chamber and the steering coil current as measured at the power supply. These are read by addressing A to D converters via a multiplexer. There are several processes within the servo control. The gantry item selects a '$2R$ value' from a look-up table, which is created during machine commissioning. The $2R$ ion chamber items are compared and provide an 'error process item'. Finally the outputs from these two internal processes together with the $2R$ current items are processed to provide the output, which is transmitted to the power supply via a multiplex unit in the control area. Adjustment of the servo characteristics, e.g. gain, offset etc, is carried out by editing 'calibration data' used in the software routines which generate the process items. While the control process is going on all the items are available for display on the VDU and are continuously monitored to ensure that they are within allowed limits, thereby contributing to the interlock systems of the accelerator.

This has been a brief description of one particular system. There are many different ways of implementing computer control but all have the same aims of improved performance, versatility and availability of information.

### 10.5.3. A novel application of computer control

The MLC developed by Philips Medical Systems—Radiotherapy makes use of a novel control system to first measure and then adjust the position of each of the eighty leaves. This makes use of optical, video and computer techniques. The optical arrangement is shown in Figure 10.8.

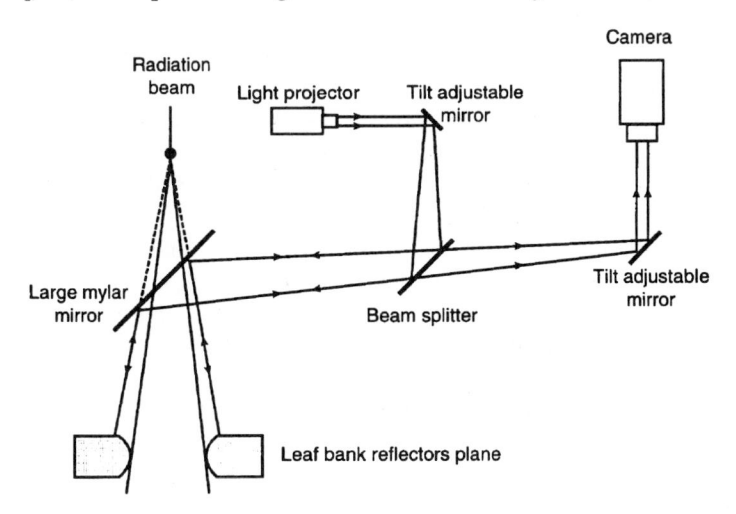

**Figure 10.8.** *The optical arrangement by which the Philips MLC is controlled.*

Each of the leaves is fitted with a reflector, which is illuminated by the main optical beam. In addition there are four fixed reflectors mounted on the collimator frame. These are then viewed by a CCD video camera via a beam splitter. The video image is captured 12.5 times per second (which is a sub-multiple of the video frame rate) and processed within the computer to identify the four reference reflectors and each of the eighty leaves. Failure to find a reflector within predefined limits of position and brightness causes the system to interrupt and for the 'MLC ready' interlock to the accelerator to be opened. If any spurious reflections are detected this also interrupts the process.

The calculation of the position of each leaf from the measured positions of each of the reflectors in the video image is based on stored calibration data. The measurement of the position of the reference reflectors provides a continuous check that the stored calibration data and the algorithms for the calculation are valid.

The measured position of each leaf is displayed on a graphic display so that the radiographer can visualize the shape of the field defined by the eighty leaves.

Movement of the leaf positions is by eighty miniature DC motors each driving a small lead screw through a nut fixed in each leaf. Control is achieved by a servo where the actual position for each leaf is compared with the prescribed position of each leaf. The leaf motors are driven to minimize these differences but will be inhibited if a collision detection algorithm predicts a collision with other nearby leaves. When all the leaves are within the tolerance band, power is removed from all the motors to prevent hunting. The measurement process continues so that if a leaf moves (or the prescription is changed) the control loop is re-energized.

The control signals to the eighty motors are multiplexed to avoid large numbers of individual conductors in connecting cables. The demultiplexer and the final drive electronics are mounted on the treatment head and therefore radiation damage to these components is an important consideration.

Prescription files for irregularly shaped beams can be generated and imported from an external treatment planning computer or can be entered from the computer terminal. Rectangular beams are pre-programmed so that entry of four parameters generates the required prescription file. These parameters can be entered from the computer terminal or varied using thumbwheel control potentiometers on the treatment head so that real time control is possible.

Although the idea of multileaf collimation pre-dated the integration of computer control into linear accelerator systems widespread application of these devices became possible only after reliable computer control systems were introduced.

## 10.6. CONTROL CONSOLES

Most of the circuits and assemblies described in earlier chapters are located inside the treatment room with some mounted on the rotating structure of the gantry. They have to be monitored and their operation adjusted when the system is producing radiation. The monitoring and adjusting systems have thus to be wired out to a control console outside the treatment room.

There are two distinct types of control operation: the first when the machine is being maintained, calibrated and adjusted, when access to the more fundamental machine operating parameters is required; the second when the machine is being operated for the treatment of patients when control of dose and all the other prescribed parameters is the prime concern. Many machines have separate control consoles for each of these functions, both being situated in the control area immediately outside the treatment room.

The machine, or engineer's, control unit, would incorporate circuits to adjust and monitor various functions including vacuum pressure, the various power supply units and field flatness, for example. Facilities for connecting an oscilloscope to observe various waveforms, e.g. magnetron high-voltage pulses, are also required in this unit. The functions dealt with by the machine control unit are required for setting up the machine, for servicing, for fault finding and for monitoring for record purposes. They are not required in routine operation of the equipment for radiation therapy once the system has been properly set up.

The treatment control unit is used to prepare for and to monitor the irradiation of patients. The main functions carried out on this unit are setting up the dose monitor, selecting the treatment mode to be used, the various 'confirm' actions as already discussed and observation of the dose, dose rate and any other parameters that need to be monitored as treatment progresses.

The main controls for moving the mechanical systems, as discussed in chapter 9, are inside the treatment room so that the patient is under direct observation when the machine is being set up for a treatment. However for rotation therapy, or when automatic set-up is being used, some or all of the movement controls need to be duplicated on the treatment control unit.

The machine control unit and the treatment control unit can be, but are not always, incorporated into the same console panel, although they serve quite different purposes. Some current linear accelerators use an array of push buttons, potentiometer knobs and rotary switches to control and select the functions described, and present information, for example, monitor doses, in the form of digital read-outs. A photograph of a unit of this type is shown in figure 10.9(a). However, for most currently produced linear accelerators, all the control and read-out functions

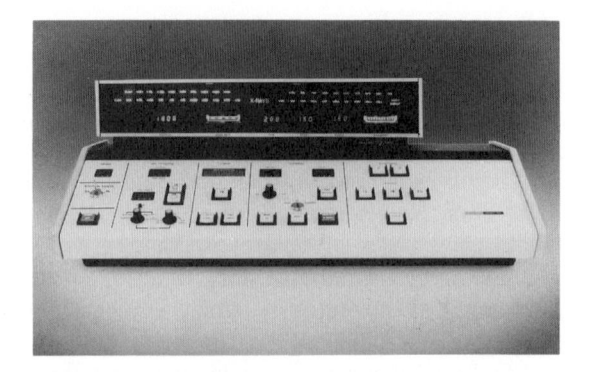

(a)

(b)

**Figure 10.9.** *(a) A control unit requiring manual entry of machine settings. (Courtesy of Varien-Tem.) (b) A control unit using a PC as a data entry terminal. (Courtesy of Elekta Oncology Systems.)*

are handled on a keyboard and VDU. Here the information from the control and read-out circuits has to be interfaced and coded into the control unit via a microprocessor or will have been fully integrated as described above. A photograph of a unit of this type is shown in figure 10.9(b). The system might also include a bar code reader for identifying a coded treatment sheet as part of an automatic set-up system and a printer, which records the details of each treatment on the treatment sheet.

## 10.7. GENERAL COMMENT

It is probably true to say that the development of control systems is the current 'leading edge' in linear accelerator technology. In most systems the internal monitoring, many interlock functions, the machine control operations and the automatic control and recording of machine and patient parameters are all handled by computer. Although the performance of accelerators is undoubtedly enhanced by computer control the management of the large number of data is a task that should not be under-estimated. The data for each patient have to be entered into the machine's patient database. This can be carried out at a remote terminal, if such a device is provided, thus avoiding the conflict between the time required for data preparation and treatment delivery. Regular back-up of data files together with other housekeeping tasks requires a significant commitment of time. Connectivity with other radiotherapy equipment such as simulators and treatment planning systems is becoming more important as the number of data required to specify an individual treatment increases. Network connection between machines, to allow transfer of patients, and to hospital administration systems, to allow for scheduling of appointments, is becoming a common requirement. These functions, while not being within the traditional ambit of radiotherapy physics and engineering, present challenging problems in security and safety.

# CHAPTER 11

# TREATMENT VERIFICATION

Verification that each treatment is delivered as intended is necessary because of the very tight tolerances imposed by steep dose response curves for both tumours and normal critical organs. It can be considered as the final stage in the treatment control process. If the treatment delivered is verified then all that is required is a record that this is the case but in the event of detection of a discrepancy between intended and actual treatment the verification process becomes part of the feedback mechanism by which discrepancies can be rectified.

Treatments can be verified at three levels to determine whether the machine settings are such as to deliver the prescribed treatment, whether the correct dose is delivered with those machine settings as intended and whether that dose is delivered to the correct parts of the patient.

## 11.1. VERIFICATION OF MACHINE OPERATING CONDITIONS

In the previous chapter the link between the treatment prescription for an individual patient and the machine settings, or treatment parameters, necessary to deliver that treatment has been discussed. Table 10.2, while not comprehensive, gives examples of the machine settings that have to be made to fulfil certain parts of the prescription, e.g. the prescription of field size is delivered by appropriate setting of the collimators. Table 11.1 gives the minimum list of treatment parameters that might be expected to be verified on a dual-mode machine; mis-setting of any of these parameters would lead to the treatment being delivered incorrectly. Which of these parameters can be precisely defined (P) and which can only realistically be defined within a range of tolerance (T) are also given. Some of the entries on the list correspond to a single parameter but others correspond to multiple parameters. The field size requires a minimum of two parameters for a simple symmetric collimator. This increases to four parameters for asymmetric, or independent collimators and up to eighty-four parameters

**Table 11.1.** *Machine parameters that should be verified to ensure acceptable treatment.*

| Machine parameter | Precision required P | Tolerance allowed T |
|---|---|---|
| Mode of treatment (x-rays or electrons) | P | |
| Beam energy | P | |
| Monitor units | P | |
| Wedge filter | P | |
| Field size | | T |
| Collimator angle | | T |
| Gantry angle (for all treatments) | | T |
| Gantry stop angle (for rotation treatments) | | T |
| Couch height | | T |
| Couch horizontal positions | | T |
| Couch angles | | T |
| Accessories e.g. shadow tray | P | |
| Treatment time | | T |

to describe the field size produced by an MLC such as that described in chapter 6.

On older machines select and confirm systems are, in principle, verification systems, the verification step being taken by the operator. In practice modern verification systems are computerized and often linked with automatic recording of treatment delivery and provide one of the treatment interlocks. An interlock is closed when all the parameters to be verified are within tolerance. Computerization has allowed for the range of verified parameters to be increased and has allowed a degree of flexibility in setting of tolerance levels, which can be customized for different types of treatment (Rosenbloom *et al* 1977). The tolerance levels chosen need careful consideration, taking into account on one hand what is achievable and on the other hand what are the consequences of failure to set each parameter accurately.

The tolerance achievable for any parameter cannot be guaranteed to be better than the manufacturer's specification for reproducibility. In the case of field size IEC 977 (1989b) recommends that the reproducibility should be within ±1 mm (to which could be added the maximum deviation between the indicated field size setting and the actual field size, which may be a further ±3 mm). A clinical situation where maximum geometric accuracy is required is a high-dose treatment to the head and neck region near to part of the central nervous system. In this case it might be reasonable to set the verification tolerance to ±1 mm. In other situations accuracy is unavoidably compromised by movement of internal organs, for example in the pelvis or thorax. In these cases a tolerance of ±5 mm might be more appropriate. In these examples the tolerance is determined by clinical factors but in some cases tolerances will be determined by the treatment technique employed.

In the head and neck treatment requiring tight tolerance on field size there might be considerable day to day variation in gantry and collimator angle when the beam is set up to align with entrance and exit marks on a treatment shell. The variation will be necessary to allow for the fact that the patient might not lie in exactly the same position each day. In this case the machine verification tolerances would have to be set quite wide, maybe $\pm 5°$.

Widest tolerances are usually used for couch positions because a patient cannot be assumed to lie in the same place on the couch on each day of treatment.

If the patient is immobilized by some form of rigid attachment to the treatment couch then it becomes practical to use tighter tolerances on all mechanical settings. Immobilization is used in an attempt to ensure that this precision is not compromised by patient movement. It should be remembered that total immobilization is not possible, except perhaps in the extreme example of stereotactic radiosurgery discussed in chapter 16, and that it is ineffective in controlling the movement of many internal organs.

Verification systems can be integrated in the linear accelerator's control system or can be a separate unit interfaced to the accelerator's control system. In either case, for each individual patient, it is necessary to enter and store the values of all verified parameters for subsequent verification. It is also necessary to be able to edit these values should the requirements change during the course of treatment and to be able to override any item which is out of tolerance if this becomes necessary and can be justified.

## 11.2. VERIFICATION OF DOSE DELIVERY

Verification of the dose delivered during treatment by *in vivo* dosimetry is one of the quality assurance procedures recommended in the WHO publication (1988) on quality assurance (QA) in radiotherapy. It is possible to estimate the dose delivered to the patient, based on the known performance of the accelerator, the modelling of dose distribution in the treatment planning computer and the anatomical details of the patient from diagnostic imaging techniques. Unfortunately the estimate is subject to many uncertainties, including the accuracy in setting the machine for each patient for whom treatment has been prescribed. Direct measurement is independent of all previous assumptions and is therefore useful in verification of treatment precision, although, as will be shown, the measurement techniques available are themselves only capable of moderate precision.

The applications of *in vivo* dosimetry are the following.

(i) *Verification of particular techniques.* When a treatment technique is first introduced it is likely that the factors used will be calculated and if appropriate the treatment will be modelled in a treatment planning computer. There might be some measurements in a phantom to confirm that the technique results in the correct dose delivery. Having taken these precautions it is still useful to measure and document the doses actually received by the first few patients to be treated.

(ii) *Verification of prescribed dose to individual patients.* The prescribed dose is usually specified at the centre of the target volume, which is not often accessible to dosimeters. Although not quite direct the prescribed dose can be verified by measuring the dose at either the entrance to, or exit from, each beam, the measurement then being compared with the treatment plan. Measurement at the entrance requires the dosimeter to be covered with a sufficient thickness of build-up material which obviously compromises skin sparing during the measurement. This is not significant if the measurement is carried out during just a few fractions of a fractionated treatment, but would be inappropriate if measurements were required during the entire course of treatment.

Measurement on the exit surface avoids this problem but precision in comparison with the expected dose is reduced as calculation of dose at the exit surface is somewhat uncertain.

(iii) *Verification that the dose to a shielded structure is acceptably low.* The most obvious example of this application is the measurement of eye dose during a treatment of a tumour in the head and neck region. Many radiotherapy departments carry out such measurements routinely. Doses near to the edge of treatment beams and in the shadow of shielding blocks are very difficult to predict as the contribution from scatter, both from the collimators and from the patient results in steep dose gradients. These dose gradients also have to be taken into account in the performance and interpretation of measurements.

The methods most commonly available for *in vivo* dosimetry are thermoluminescent dosimetry (TLD) and the use of silicon diodes. A review of *in vivo* dosimetry is given in chapter 10 of the work of Williams and Thwaites (1993).

*11.2.1. Thermoluminescent dosimetry*

When thermoluminescent materials absorb dose from ionizing radiation, some of the energy is stored at trapping centres within the crystalline structure. This stored energy is then released in the form of light by heating the material to a high temperature. The light is measured by a photomultiplier and can be related to the absorbed dose. Lithium fluoride (LiF) is the most common TLD material for radiotherapy dosimetry. It is available in several physical forms including powder, sintered pellets and in combination with Teflon as thin discs. Dosimeters with dimensions of a few millimetres allow for measurements with adequate sensitivity and spatial resolution. The material has a mean atomic number (8.1) close to that of soft tissue (7.4) and this results in low energy dependence of the sensitivity.

Dosimeters are usually used in batches so that the signal from each individual dosimeter can be compared with others from the same batch which have been irradiated to known doses for calibration. As LiF exhibits supralinearity at doses in the range that might be measured in *in vivo* dosimetry it is necessary to ensure either that there are sufficient dose levels used in the calibration or that corrections are made for this non-linearity. With care TLDs can give a precision of about ±5% standard deviation. The standard error of the mean of a group of $N$ dosimeters is of course reduced by a factor $\sqrt{N-1}$. The main disadvantage of TLD is that the measurement is not instantaneous: the dosimeters have to be placed on the patient, irradiated, removed and then taken to a TLD reader for measurement.

### 11.2.2. Silicon diode dosimetry

Semiconductor diodes can be used as radiation detectors. They produce a current proportional to dose rate when they are used in short-circuit mode, that is without a polarizing voltage applied. Although both n-type and p-type diodes can be used the p-type are preferred because they suffer from smaller changes in sensitivity as a result of accumulated radiation damage than do n-type devices. The measuring circuit is extremely simple, consisting of an electrometer with a low input offset voltage (ensuring that the bias voltage across the diode is held at zero). There is no need for particularly high gain as the current is very much higher than would be generated by an ionization chamber used for machine calibration or QA. The higher sensitivity of a diode compared with an air filled ionization chamber comes from the high density of silicon compared with air. Silicon is far from tissue equivalent with an atomic number of 14. The energy dependence of sensitivity is therefore higher than for TLDs but this is not too much of a problem in the megavoltage energy range. It is somewhat surprising that diodes for *in vivo* dosimetry are often encapsulated in stainless steel to provide the necessary build-up without unduly increasing the size of the detector.

The simplicity of the instrumentation and the fact that the reading is available instantly makes the silicon diode system a very useful tool for *in vivo* dosimetry. Multichannel electrometers make it possible to make simultaneous measurements of the dose at more than one point. Care has to be taken to calibrate the diodes at a temperature near to the temperature at which they will be used. Both the dark current and sensitivity are temperature dependent and the latter is somewhat dependent on accumulated dose.

## 11.3. VERIFICATION OF IRRADIATION GEOMETRY—PORTAL IMAGING

Although certain aspects of irradiation geometry can be verified by ensuring that the treatment machine is correctly set up, portal imaging can be used to verify the position of the radiation fields with respect to the patient's anatomy. This can include verification that field modification devices, such as shielding blocks, are correctly positioned. Portal imaging, a fuller review of which is given in chapter 6 of the book by Webb (1993), involves taking a radiographic image with the treatment beam. This can then be compared with a reference image from a simulator or a digitally reconstructed 'simulator' image derived from CT images.

Portal images taken at megavoltage energies are very poor compared with diagnostic x-ray images. This is a result of the x-ray interaction processes in the patient and in the image receptor which are close to optimum for x-rays used in diagnostic radiology but not for megavoltage x-rays. In the patient the predominant interaction is Compton scattering. The image contrast results only from changes in density rather than from variations in atomic number. Interfaces between soft tissue, air and lung tissue can be seen relatively easily but bones are less clearly imaged. The second problem arising from Compton scattering is that at these energies the scattering is narrow angle and the image receptor cannot be shielded. This further reduces contrast and increases unsharpness.

The simplest and most commonly used method is to use slow x-ray film. This needs no special technology apart from a cassette with a thin metal screen which absorbs some of the photons passing through the patient and produces electrons which expose the film.

### 11.3.1. Electronic portal imaging devices

Several types of electron portal imaging device (EPID) are available and these have many potential advantages over radiographic film.

In all cases the image is captured in a digital form so the following points apply.

(i) It is possible to vary the contrast and brightness, within the limits set by the detector and its electronics. This can be done by a combination of optimizing the image acquisition parameters, such as the exposure time, and by post-capture processing.

(ii) It is possible to implement image processing algorithms to enhance the visualization of some structures. Contrast enhancement, edge sharpening and noise suppression are the processes most often employed.

(iii) With the appropriate software tools it is possible to make measurements directly of the image, e.g. of field placement errors. This becomes simple if the reference image is in the same electronic format so that it can be displayed side by side and key features extracted and superimposed on the portal image.

(iv) It is possible to relate the brightness of each pixel to dose at points along the projection from that pixel to the target. The relationship is somewhat indirect so the dose derived in this way can only be regarded as an estimate. However there seems to be considerable potential for using these devices to estimate exit doses for the purposes of treatment verification and perhaps to provide information for the design of individual compensators (Evans *et al* 1994, ch 8.6).

The image can be formed with a relatively low dose and is instantly available so the following apply.

(v) It is possible to take double exposures including a wide field of view as the second image without excessive dose outside the target volume. This is particularly important as it is difficult to visualize structures in the normally small-field portal image, which is of inherently poor quality.

(vi) It is possible to capture and display an image after only a small fraction of the daily treatment dose has been delivered, so, if a set of decision rules can be defined, it becomes possible to use an EPID in an interactive rather than retrospective way.

(vii) It is possible to display real time images to demonstrate movement during irradiation. The simplest example is that of the effect of respiration on field coverage in the treatment of lung tumours. This can be assessed and if necessary the treatment fields can be extended.

The devices available fall into two main categories.

*11.3.1.1. Fluorescent screen systems.* The principle of a fluorescent screen EPID is shown in figure 11.1(a). The image receptor is a fluorescent screen, mounted to intercept the exit beam from the patient. The screen is bonded to a thin metal converter plate whose prime purpose is to convert x-rays transmitted through the patient to electrons, which in turn are absorbed in the fluorescent material, forming an optical image. The secondary purpose of the metal plate is to absorb the Compton electrons emitted from the exit surface of the patient, which would otherwise blur the transmission image. The image on the fluorescent screen is viewed by a TV camera via one or more mirrors so that the camera is not directly affected (and damaged) by the x-rays. The video signal is then captured electronically. In principle images are captured at the frame rate of the TV camera but in practice it is necessary to integrate over several frames in order to avoid quantum mottle and to produce a signal higher than the dark current in the camera and its electronics. First-generation EPIDs of this type use CCD cameras, which produce an image of up to $512 \times 512$ pixels, which is digitized to eight

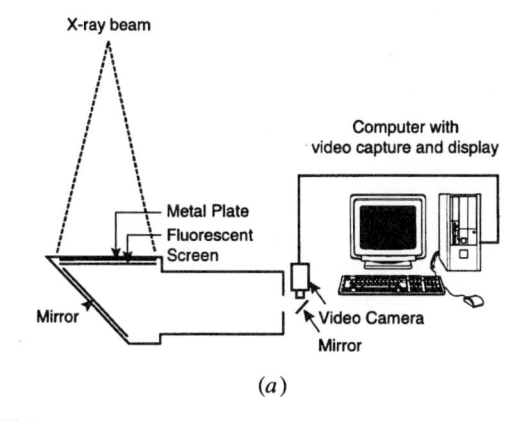

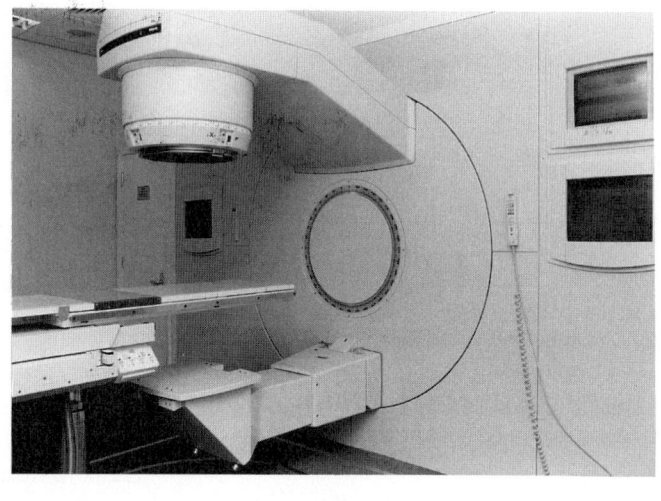

**Figure 11.1.** *(a) The principles of a fluoroscopic EPID. (b) A fluoroscopic EPID mounted on an accelerator.*

bits, that is 256 possible grey levels. Figure 11.1(b) shows a fluoroscopic system mounted on a linear accelerator. An example of an image taken with one of these devices is given in figure 11.2. Improvements in performance are possible with cameras capable of twelve- or sixteen-bit resolution but as the spatial resolution is limited to about 1 line pair mm$^{-1}$ an increase in CCD array size beyond 512 is not advantageous unless the image receptor is larger than 512 mm square.

The sensitivity of a fluorescent screen–TV camera EPID is dependent on the efficiency of all the processes involved in converting the energy carried by the x-ray leaving the patient to the final image. The materials and thickness of the metal plate and fluorescent screen affect the sensitivity,

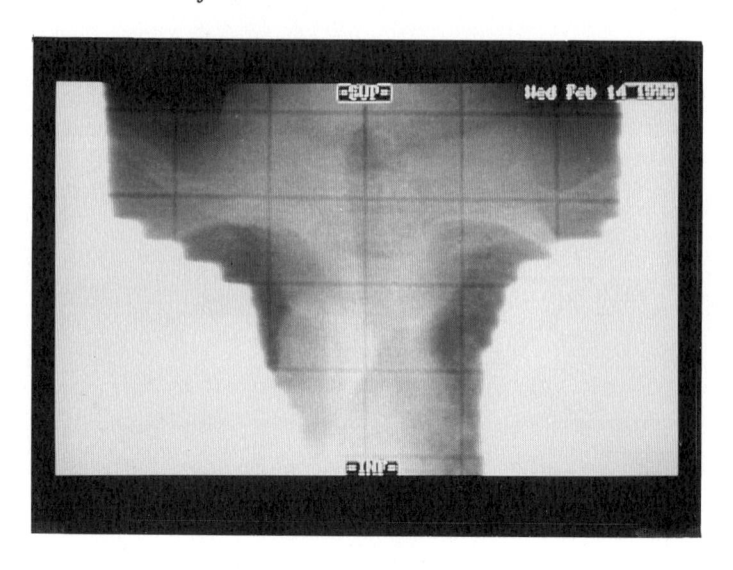

**Figure 11.2.** *A portal image taken with an EPID.*

noise and spatial resolution of the system. The response of the TV camera to the wavelength of the light emitted by the screen affects the sensitivity as does the efficiency in collecting the light. As fluorescence in the screen is isotropic only a small fraction of the light emitted falls within the aperture of the camera's lens.

There are some other options available to overcome the problem of light collection, but these are either experimental or very expensive. They include coupling the fluorescent screen to the TV camera by a large array of tapered optical fibres and forming a two-dimensional array of amorphous silicon photodiode cells in contact with the fluorescent screen.

*11.3.1.2. Arrays of radiation detectors.* These include arrays of ionization chambers and silicon diodes. The most common is the matrix ionization chamber described by van Herk and Meertens (1988) shown in schematic form in figure 11.3. This consists of an array of 256 × 256 liquid filled ionization chambers covering an area of 32 cm × 32 cm. Each ionization chamber is formed by the intersection of one of 256 collecting electrode strips with one of 256 polarizing electrode strips. The polarizing voltage of 300 V is switched sequentially to each of the polarizing electrodes. This establishes an electric field, sufficient to collect the ionization produced in the liquid, at the intersections of the polarizing electrode and each of the 256 collecting electrodes. The separation of the polarizing and collecting electrodes is 1 mm. Each collecting electrode is connected to one of 256 electrometers, which is used to measure the ionization current. The imaging sequence is therefore to sequentially measure the

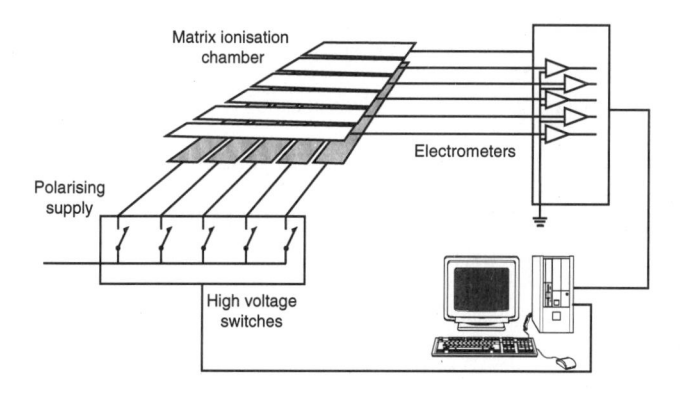

**Figure 11.3.** *Principles of a matrix ionization chamber EPID.*

ionization collected on each of the 256 collecting electrodes as each of the polarizing electrodes is energized in turn. Image acquisition time is a few seconds. The electrometers and the high-voltage switching circuits have to be mounted in the ionization chamber housing together with a small amount of control electronics. This reduces the number and complexity of interconnections between the detector and the computer system which controls image acquisition and provides the necessary facilities for viewing and manipulating the images.

The matrix ionization chamber is a two-dimensional array of detectors. An alternative is to use a linear array of detectors, which can then be mechanically scanned to produce a two-dimensional image. Such a system, described by Morton *et al* (1991), is based on a linear array of 128 zinc tungstate detectors. Although conceptually a linear array this device actually has two rows of sixty-four crystals offset by half the crystal spacing. The intensity profile from the 128 measurements is therefore smoothed but is adequately sampled to give a spatial resolution of 3 mm at the detector corresponding to 1.5 mm at the isocentre (due to magnification). The use of zinc tungstate, with a quantum efficiency of about 50% compared with about 1% for a TV based fluoroscopic system, compensates for the apparent inefficiency of mechanical scanning where only a fraction of the full field of view is sampled at any one time. The overall image acquisition time of 3–4 s is comparable with that for a fluoroscopic system.

# CHAPTER 12

# SPECIFICATION, PERFORMANCE AND CALIBRATION

The mechanical and electrical systems which make up a linear accelerator are subject to all the normal conditions relating to the specification and safety of complex mechanical, electronic and electrical equipment. This chapter, however, deals only with those items which relate directly to its use as a radiotherapy machine.

Detailed specifications for medical linear accelerators are given by the International Electrotechnical Commission (IEC 1989), and in particular part 3.1 (1989a, 1989b) gives numerical values for the mechanical and dosimetric tolerance values. Many of these have been referenced in appropriate places throughout this book. The IEC documents also give detailed methods of testing for these performance parameters. This chapter outlines a set of tests that can be used to show that the linear accelerator's performance meets its specification both at commissioning and during day to day operation. Further tests are described in chapter 14. Where there are differences between these tests and those described in the IEC standards the latter should be regarded as definitive.

## 12.1. MECHANICAL SYSTEMS

There are three basic and related specifications for the mechanical systems:

(i) the stability and symmetry of the treatment head about its axis of rotation,
(ii) the stability of the centre of rotation of the gantry and
(iii) the stability of the vertical axis of rotation of the turntable of the patient support system.

The convergence of these axes defines the position and stability of the isocentre.

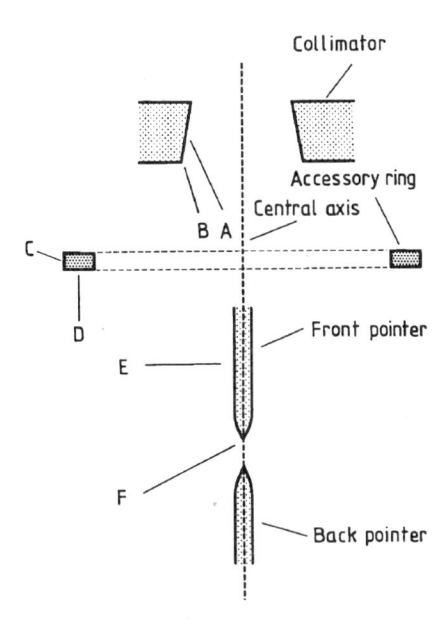

Figure 12.1. *The use of a rigid pointer in positions A–F to check centring and alignment of beam defining collimators, accessory ring and mechanical front and back pointers with respect to the central axis.*

### 12.1.1. The treatment head

The treatment head is designed to place the central axis of a radiation field with symmetrical edges on a line passing through the isocentre for all positions of the equipment. The mechanical elements which define this axis are the main bearing and the beam defining collimators (see figure 6.2).

The main bearing has to be centred on the radiation source in a plane normal to the central axis and this is usually fixed by a system of jigs or dowels on assembly while the accelerated electrons are directed onto the source by the systems discussed in chapter 5. The tolerance of the positioning of the beam defining collimators is about 2 mm and is discussed in chapter 8. However the 2 mm is an overall tolerance arising from mechanical alignment and accuracy of scales indicating the position of each collimator. Tests of symmetry of the collimators based on purely mechanical measurement, independent of the transducers and scales, should be to much tighter limits, ideally within 0.2 mm. This can be tested by placing a rigid reference pointer (supported independently of the treatment head) to touch the face of one of the collimators (see figure 12.1, position A), rotating the head through 180° about its main bearing and observing the relationship between the pointer and the other collimator. If the optical beam is illuminated, a gap of 0.2 mm between the pointer and either collimator is easily observable to an experienced eye. The test can be carried out at a higher level of precision by attaching a toolmaker's clock gauge to the pointer but this is not necessary. This test needs to be carried out on both pairs of beam defining collimators with the gantry in the vertical and both horizontal positions, and will take account of flop in the main bearing, mechanical deflections and backlash in the collimator support mechanism.

Backlash between the positions of the beam defining collimators and the 'collimator setting' read-out (which may be mechanical or electronic) can be examined by placing the rigid reference pointer at position B in figure 12.1, and observing the reading. When the scale reading is brought to this value by moving the collimators from opposite directions, the position of the collimator with respect to the reference pointer should be within 0.2 mm. (Note that this measurement is on the collimator faces: the effect of 0.2 mm backlash at this point will be magnified at the position of the patient near the isocentre. The actual setting of the scale is a separate operation related to the definition of field size, as discussed in chapter 8.)

Tests of the mechanical alignment of a multileaf collimator are extremely complex and can best be achieved by measurements of the optical and x-ray beam rather than from mechanical measurements.

Beam direction devices are normally located on the accessory ring, whose reference surfaces should be centred on the axis of rotation at manufacture. This can be tested to the same tolerance as the collimator centring using a rigid pointer in relation to the radial reference surface (position C in figure 12.1) and the surface normal to the axis (position D) when the head is rotated through 360° about its main bearing.

The mechanical front pointer is normally mounted on the accessory ring. Reference pointers, in positions E and F, can then be used to test that the pointer is centred in line with the central axis when the head is rotated about its main bearing. If the pointer is designed for use over a range of SSDs it should be tested over the full range, with the tightest tolerance being applied to the most commonly used SSD. The front pointer mounting may be placed at any position on the accessory ring so, if the mounting is correctly made, then, at this stage of the test procedure, the front pointer should centre correctly at any position. This will only happen if the front pointer mounting is properly set up and can be tested by placing the mount at four diametrical points on the mounting and repeating the observation at positions D and E in figure 12.1. The front pointer and its mounting form a less rigid structure than the main treatment head and so a reasonable tolerance on these tests is ±0.5 mm.

The front pointer may now be used as a reference for lining up the optical beam pointers and the main optical beam. The front pointer is used to centre a sheet of graph paper in the plane of the isocentre for this purpose. The back pointer can be adequately lined up with the front pointer by eye, again to a tolerance of ±0.5 mm.

## 12.1.2. The gantry

The stability of the isocentre can now be tested by using the centred front pointer, which is set with its end nominally 1 m from the radiation source at the approximate position of the isocentre.

The rigid reference pointer, supported independently of the gantry, is mounted at position F in figure 12.1 and the relationship between the two pointers is observed as the gantry is rotated through 360°. Alternatively, a sheet of graph paper is mounted in the vertical plane normal to the axis of the rotation of the gantry and touching the front pointer. The position of the end of the front pointer is marked on the paper for a number of gantry angles as the system is rotated through 360°. These points should form a closed loop (not necessarily circular) as the system deflects and the major diameter of this loop should not exceed 4 mm. This information can be used to make a final adjustment of the position of the end of the front pointer, to be at the centre of symmetry of the loop which is the isocentre. This may require the SSD scale to be reset so that the indication at the isocentre is exactly 100 cm.

## 12.1.3. The patient support system

The axis of rotation of the turntable of the patient support system should pass through the isocentre within a tolerance of $\pm 2$ mm. The procedure under 12.1.2 above places the end of the front pointer at the isocentre. The stability of the centre of rotation can be examined by placing a sheet of graph paper on the couch in contact with the front pointer and marking the positions of the front pointer as the couch is rotated through 360°. The locus of these points should be within a 4 mm diameter circle. The couch will deflect under the weight of a patient, so this test should be performed on a loaded couch.

To check that the axis of rotation of the turntable is vertical, a plumb bob is suspended from the front pointer holder in such a way that its end is initially at the position of the isocentre and with the central axis directed vertically downward. Its position is marked on paper on the couch. The couch is then taken to its lowest position, the plumb bob lowered to come into contact with the paper, and its position marked on the paper as the couch is rotated. The points marked again should be inside a 4 mm diameter circle. If the couch lift is truly vertical and the gantry is rotated so that the central axis of the collimator system is also vertical it should be possible to repeat the test of isocentric couch rotation at any couch height.

## 12.1.4. Note on isocentric accuracy

An accuracy of $\pm 2$ mm, corresponding to the tip of the front pointer being within a 4 mm diameter sphere centred on the isocentre for all combinations of collimator gantry and couch rotation, may seem imprecise when mechanical measurements and manufacturing can be carried out to tolerances of at least an order of magnitude better than this. However in engineering terms this is still a difficult tolerance to achieve for the movement of such a heavy structure. Fortunately the required tolerance is in most cases also limited for anatomical and physiological reasons. The object of the design of the isocentric system is to place this point in space either at the skin entry point, or at the position of the centre of the volume to be treated. Because of physiological movement of an immobilized patient and the limits of reproducing patient positioning these points cannot be defined to an accuracy better than ±2 mm unless extreme measures are taken as in the case of stereotactic radiotherapy, which will be discussed in chapter 16.

### 12.1.5. Movement control

The only movement whose speed is critical is the rate of rotation of the gantry when it is being used for moving beam therapy. For this purpose the speed is conveniently expressed in degrees per monitor unit (where one monitor unit corresponds to a defined dose at a reference point) and it can be checked by a measurement of angular movement during irradiation in moving beam mode of operation.

The mechanical tests to be performed on a linear accelerator are discussed in more detail in *IPSM Report* 54 (1988).

## 12.2. RADIATION OUTPUT

Examination of the radiation output from a linear accelerator involves, in principle, the two types of measurement illustrated in figure 12.2. Figure 12.2(a) shows a measurement in air where the dose sensitive detector (or dosimeter probe) may be inside the useful beam, as shown, or at any point outside the beam where stray or unwanted radiation has to be detected and measured.

The most commonly used radiation detector is an ionization chamber, whose sensitivity depends on its volume, i.e. the volume of air in which the ionization produced by the radiation is collected and measured. Inside the radiation beam, and immediately adjacent to its edge, the dose will vary rapidly with position and so a small detector is required. In this region high dose rates are available, so the conflicting requirements of a maximum-volume detector to give an easily measured ionization signal, and a small-volume detector to give good spatial resolution, can be satisfied by using a detector of not more than about 5 mm in diameter. Outside the edge of the beam, the system is designed to give a very low dose rate, but spatial variations in dose rate are likely to be slow, so here the conflicting

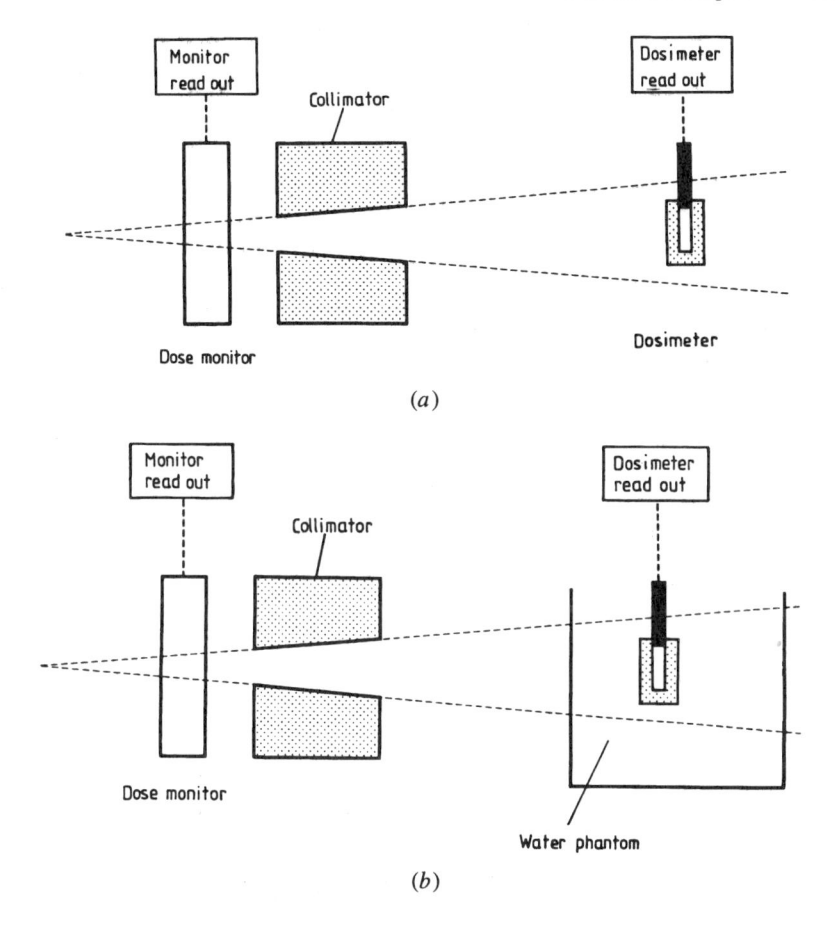

**Figure 12.2.** *The use of a dosimeter probe to make measurements (a) in air and (b) in a phantom.*

requirements can be met by using an ionization chamber several centimetres in diameter.

For x-ray measurements in air, the air volume in the ionization chamber needs to be surrounded by a thickness of low-atomic-number material (usually plastic or carbon), sufficient to give 'electronic equilibrium'.

For x-ray measurements in a scattering medium, as illustrated in figure 12.2(b), the medium itself will provide the material for secondary electrons to establish electronic equilibrium. In principle, a thin-walled chamber (thin in relation to the secondary-electron range in the wall material) will be in electronic equilibrium at depths in the medium greater than that for the peak of the depth dose curve (figure 12.3). The considerations concerning detector size and spatial resolving power already discussed still apply in a scattering medium. In a water phantom the ionization chamber needs to be

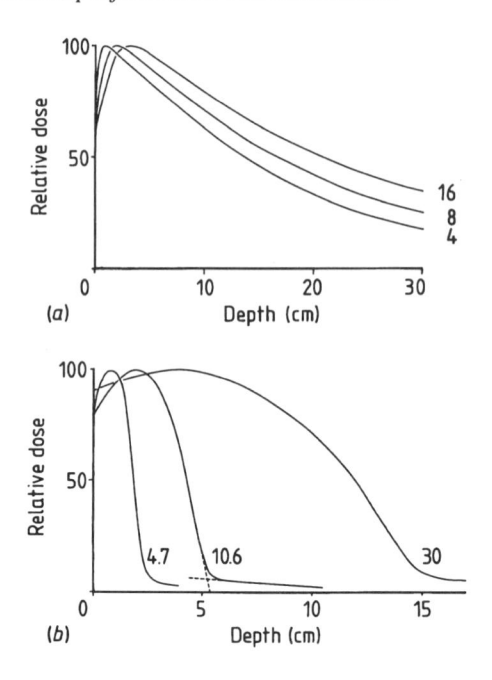

**Figure 12.3.** (a) Central axis depth dose curves for 4, 8 and 16 MV x-rays, for a 10 cm × 10 cm field at 100 cm SSD. (b) Central axis depth dose curves for 4.7, 10.6 and 30 MeV electron beams for a 10 cm × 10 cm field.

surrounded by a waterproof sheath if its basic construction does not make it waterproof. In the special case where the ionization chamber has been calibrated to measure absorbed dose it should be surrounded by the plastic cap with which it was calibrated even when used in a scattering medium.

When the linear accelerator is being used as an electron beam generator, the two basic types of dose measurement illustrated in figure 12.2 are still required. The signal in the ionization chamber detector is now produced by fast electrons from the beam penetrating the walls of the ionization chamber and ionizing air inside it. The walls of the ionization chamber detector have to be thin in relation to the electron range to allow this penetration, and the whole detector needs to be thin in relation to the range of the electrons in water if it is to be able to resolve the electron beam depth dose curve. These conditions are easily met for electrons in the range considered in this book.

Detailed discussions of the principles and practice of radiation dosimetry may be found in more specialised publications (Greening 1985, ICRU 1969, 1972, HPA 1983, Lillicrap et al 1990), the last reference being to the UK code of practice for high-energy photon therapy dosimetry based on the NPL absorbed dose calibration service.

## 12.2.1. Specification of the dose monitor

The dose monitoring system is described in chapter 7 and can be tested by the use of the arrangements illustrated in figure 12.2. The system needs to be specified in terms of sensitivity, precision and linearity.

### 12.2.1.1. Sensitivity.
For radiotherapy treatments the sensitivity of the dose monitoring system has to be such that one monitor unit corresponds to a dose of 1 cGy in water at 1 m from the radiation source. The detailed relationship between the reading in monitor units and the absorbed dose is a function of field size and this will be discussed later. The specified sensitivity can then apply precisely at only one field size and this is an arbitrary choice. A field size of $10 \times 10$ cm$^2$ at 1 m SSD is often used as the 'reference field' for which the system is calibrated.

As discussed in chapter 7 the response of the monitor ionization chamber is, in principle, dependent on dose rate and the specification of sensitivity needs to take account of this. The specified sensitivity can then only apply to one dose rate but it is practical to design the monitor ionization chamber to be independent of dose rate to better than 0.5% over the range used in clinical practice.

The above discussion applies to both the channels for monitoring dose. The dose rate monitor should have a sensitivity of one monitor unit per minute for a dose rate 1 cGy min$^{-1}$ for reference field conditions and is subject to the variations with field size and dose rate already discussed.

### 12.2.1.2. Precision.
Precision needs to be considered in terms of short- and long-term precision.

All radiation measurements in the field and outside the field use the monitor dosimeter as the reference system; that is, all results are determined as measured dose per monitor unit. Short-term precision can be considered in relation to groups of such measurements and can be specified in terms of the arrangement in figure 12.2(b), where the phantom is at 1 m from the radiation source and the dosimeter probe on the central axis. It is quite practicable to specify the precision of this type of measurement as 0.2% for the standard error on 10 measurements of the ratio of the dosimeter read-out to the monitor read-out, where the dosimeter read-out corresponds to an absorbed dose of about 1 Gy. This requires that the test dosimeter (dosimeter probe and read-out system) has a rather better precision than that just stated. It can be tested to any required degree of accuracy by exposing the dosimeter probe to the radiation from a radioactive source in well defined geometry and using a stop watch to determine equal increments of dose.

Long-term precision extends over the lifetime of the equipment and needs to be tested regularly. The system as a whole is subject to long-term drifts. While, for any particular test, it is practicable to expect the precision specified in the previous paragraph, long-term effects will cause the mean values to vary by more than this amount. A reasonable long-term specification is for the mean values to stay within ±2%. This is dependent not only on the operation of the beam monitoring system itself, but also on the operation of all the systems which stabilize the energy of the accelerated electrons and the position of the radiation source. The system will require periodic readjustment to stay within the specified limit of ±2%. Again, long-term precision requires long-term constancy of the probe dosimeter, which can be checked by the use of a long-half-life radioactive source, as discussed above, and correcting for half-life. A strontium-90 source is often used for this purpose, having a suitable half-life (30 years).

This discussion has been about dose measurements and their relationship to the readings of the dose monitor. An exactly parallel discussion and associated specifications could be applied to dose rate measurements and dose monitoring.

*12.2.1.3. Linearity.* The previous two sections concerning the sensitivity and precision of the dose and dose rate monitors apply at a single point on the scale, that corresponding to a dose or dose rate in water, of about 1 Gy or 1 Gy min$^{-1}$ respectively at 1 m. It is also necessary to specify the properties of the system at all points on the scale or read-out, and this requires a statement of linearity. The relationship between monitor dose and measured dose, under the conditions of figure 12.2(b), for a fixed field size, should be linear within 1%. In other words the ratio of measured dose and dose monitor readings should be within 1% at all monitor scale readings. Dose monitors with digital read-out can meet or better this specification.

## 12.3. PROPERTIES OF RADIATION BEAMS

The properties of the radiation beams can be measured in a water phantom, as in figure 12.2(b), water being regarded as a suitable substitute for soft tissue. The variation of dose with depth, or the measurement of the dose distribution of the whole field, can be determined by making a sufficient number of measurements with the dosimeter probe in different positions. In practice this process can be automated by mechanically scanning the dosimeter probe in the water phantom and recording the resulting dose measurements.

*12.3.1. X-ray and electron depth dose curves*

The results of measurements of this type can be summarized in figure 12.3(a), which shows depth dose curves for x-rays, and figure 12.3(b), for electron beams.

In figure 12.3(a) the curves are labelled in megavolts, which is a relatively crude method of specifying radiation quality. For example, '8 MV' is shorthand for 'that continuous x-ray spectrum which is generated when 8 MeV electrons are incident on a target'. This spectrum will extend up to a photon energy of 8 MeV, but its detailed shape will depend on the type of target used, as well as on subsequent filtration. These points have already been discussed in chapter 6. In addition to the filtered radiation from the x-ray target, the beam incident on the water phantom (figure 12.2(b)) will also include scattered radiation from the beam defining system.

The factors which determine the penetrating properties of the x-rays generated by electrons with energy in the megavoltage range are discussed in detail in the *British Journal of Radiology* (1996), which suggests that penetration in water is itself the best measure of radiation quality and recommends the use of a quality called $D_{10}$, the percentage dose at a depth of 10 cm. Since the detailed shape of x-ray depth dose curves as in figure 12.3(a) depends on field size and SSD, the definition of $D_{10}$ has also to specify these quantities, the chosen values being a field size of $10 \times 10$ cm$^2$ and an SSD of 1 m.

In common with the fields from any generator producing x-rays in this quality range, the depth dose curves have maxima at increasing depths below the surface as the energy of the accelerated electrons is increased.

In figure 12.3(b) the electron depth dose curves have been labelled in MeV, the energy of the accelerated electrons. This again is a crude description because the energy of the electrons arriving at the surface of the water phantom will also depend on the scattering foil and on the presence of scattered electrons from the applicator. The *British Journal of Radiology* (1996) also suggests the use of penetration in water as the relevant measure of radiation quality and the use of $d_{80}$ as the reference parameter: this is the depth of the 80% isodose on the central axis. Although the electron beam depth dose curves show a limited penetration in water, they do not reduce to zero dose at depth but have a characteristic 'tail' beyond the depth at which they can be extrapolated to zero. This tail is produced by an unwanted x-ray component in the field. The proportion of the dose due to x-rays is given by the intercept of the extrapolation of the tail and the falling depth dose curve (see figure 12.3). Since the x-ray component is unwanted radiation, a limit to its value has to be specified at between 5 and 10% for the lower and upper end of the electron energy range.

In summary, the radiation quality of the x-rays, or electron beams generated by a linear accelerator, should be specified as $D_{10}$ and $d_{80}$ statements.

### 12.3.2. Isodose charts

More detailed information about the radiation beams is provided by isodose charts, examples of which are given in figure 12.4. An isodose chart normally gives the dose distribution in the central plane of the radiation

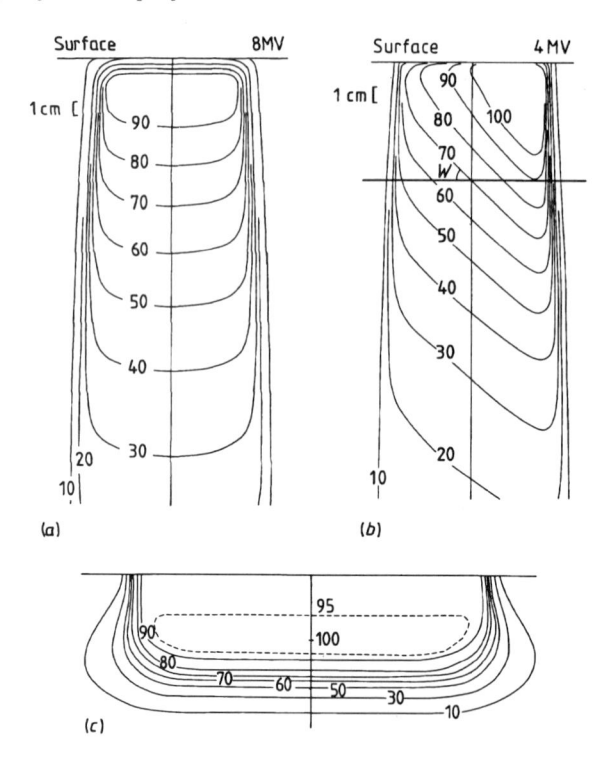

**Figure 12.4.** (a) An isodose chart for 8 MV x-rays for a 10 cm × 10 cm field at 100 cm SSD. (b) A wedged isodose chart for the same conditions as in (a); the angle W is the wedge angle. (c) An isodose chart for an 8 MeV electron beam for a 10 cm × 10 cm field defined by an electron collimator (figure 6.12).

field. The field, of course, provides a three-dimensional dose distribution and this further information can be handled by using isodose charts for planes parallel or perpendicular to those shown.

The special features of the x-ray isodose distributions provided by linear accelerators are the well defined beam edges (narrow penumbra), which are mainly a consequence of the small focal spot size, the 'flatness' of the field, i.e. the uniform dose distribution across the field, the stability of the dose distribution, the ability to produce any desired type of wedged field (see figure 12.4(b)) and the ability to provide very large fields.

Because the dose at any point in the field includes a significant component of scattered radiation from other parts of the field, the isodose lines inside the beam tend to become less flat with increasing depth. In other words the field can only be truly flat at one depth, and this depth can be chosen when the beam flattening filter is designed. If this depth is greater than the depth for peak dose, then the field has to be 'over-flattened' on the anterior side

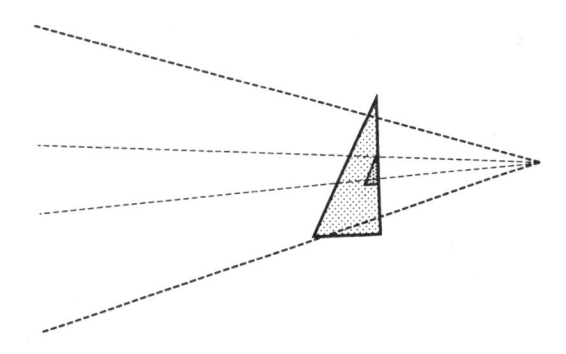

**Figure 12.5.** *A diagram showing a cross section of wedge filters to cover large and small fields.*

of this depth.

The dose distribution described by the isodose charts has to be the same for all positions of the gantry and remain constant with time. This stability depends on the operation of the feedback systems, described in chapter 5, to control the energy of the accelerated electrons and the position of the focal spot. The stability of the x-ray field needs to be specified, both in terms of the dose distribution across the field, and in terms of the depth dose values. The required dose distribution across the field is normally specified as staying within 3% at the depth in a phantom corresponding to peak dose, for the largest field available. This may seem rather a wide tolerance in relation to the other values quoted but it is realistic in terms of the complex factors which control the field flatness.

The depth dose stability can be expressed in terms of $D_{10}$. A tolerance of 1% of the maximum dose, i.e. from 70 to 71%, is a reasonable specification.

The special properties of a wedged field can be specified by the use of a wedge angle as defined in figure 12.4(b). The stability of this field is determined by the same factors as those that affect an unwedged field and by the mechanical stability of the position in which the wedge filter is held.

Wedge filters are often designed to be used up to a specified field size. As the field to be wedged is increased in size, the thickness of the wedge filter at the position of the central axis is increased, as is the loss in dose rate on the central axis when the wedge is in use. Since the wedge filters are normally used for small-field treatments there is no need to accept the very large drop in dose rate that would be required if wedge filters were used to cover the largest available fields (see figure 12.5). Alternative methods of producing wedge fields have been described in section 6.4.

Figure 12.4(c) shows an electron beam isodose chart. The important features of this are the limited depth to which the field penetrates and the poorly defined beam edge, produced by side scattered electrons. This field is also specified in terms of dose uniformity across the field (tolerance ±3%)

and in terms of $d_{80}$, the relative dose at this depth, to be within $\pm 1\%$.

The properties and application of the radiation fields are discussed in more general textbooks on radiotherapy physics (Meredith and Massey 1972, Johns and Cunningham 1983, Mould 1985, Williams and Thwaites 1993) and in ICRU reports (1969, 1972).

## 12.4. CALIBRATION OF THE DOSE MONITOR

The techniques of measurement and the data discussed in the previous two sections all depend on relative measurements. The results of these measurements can be used to determine dose in defined units at any point in the radiation field by performing a calibration at a suitable position in the field. This requires the use of a calibrated instrument, usually an ionization chamber. The basic arrangement for such a measurement is as shown in figure 12.2(b). For x-rays the recommended depth of the calibration is 5 cm for energies up to 10 MV and 7 cm for energies above 10 MV. Although any measurement of dose can be regarded as a calibration, the term 'definitive calibration' has come into use. 'Procedures for the definitive calibration of radiotherapy equipment' were set out by the Institute of Physical Scientists in Medicine in 1992 and are reproduced in HSE PM77 (1992). Definitive calibration describes the initial calibration of a linear accelerator or the re-calibration after a critical component, such as the dose monitor, has been replaced. The definitive calibration requires extreme care in both measurement and recording and should include at least two independent measurements carried out by different people, ideally with different dosimeters, although usually there will be a common link at the level of the secondary standard. Consistency between radiotherapy centres is achieved by following an appropriate code of practice or dosimetry protocol. Sections 12.4.1 and 12.4.3 refer specifically to UK codes of practice; some others in common use are listed in table 12.1 following section 12.4.3. Inter-departmental audits provide a further valuable check that the dosimetry standards have been correctly disseminated from the national standardizing laboratory.

This procedure can be used to calibrate the dose monitor and is usually used to adjust the sensitivity of the system so that one monitor unit corresponds to 1 cGy on the central axis at the depth dose curve maximum for a $10 \times 10$ cm$^2$ field at an SSD of 1 m. A typical radiotherapy dose, for a fractionated treatment, is of the order of 1 Gy. Since these doses can only be measured to the uncertainties mentioned in the previous paragraph, it would be false accuracy to calibrate the dosimeter in smaller units than 1 cGy. The dose in the phantom includes a component due to scattered radiation, which is a function of field size. The dose per monitor unit will then be less than unity for smaller fields and greater for larger ones as shown in figure 12.6. This curve needs to be measured for each radiation quality on any particular

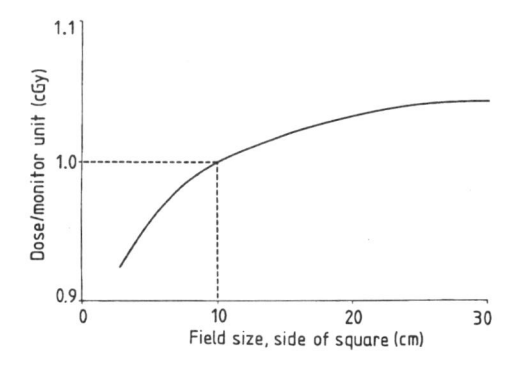

**Figure 12.6.** *Dose per monitor unit at the depth of the maximum dose as a function of field size at 100 cm SSD for 8 MeV x-rays.*

machine. It is given for square fields in figure 12.6, but, for a given field area, a long narrow field will contribute less scattered radiation at the field centre than a square field, i.e. the dose per monitor unit is smaller. This can be taken into account for any particular x-ray machine by determining the equivalent square field size for each rectangular field (*British Journal of Radiology* 1996) or by use of measured 'elongation factors'. These give the ratio of the dose at the centre of a rectangular field to that for a square field of the same area, as a function of the ratio of the lengths of the long and short axes of the field. In general for ratios up to 3:1 of the long and short length axes, the reduction of dose per monitor unit is likely to be less than 3% for megavoltage x-rays. For irregularly shaped fields, the dose per monitor unit is not likely to differ from that for a square field of the same area by more than 5%. To determine the dose per monitor unit for irregular fields to an accuracy of 2% requires relatively complex calculations (Cunningham *et al* 1972, Hounsell and Wilkinson 1990).

*12.4.1. X-ray calibration standardization*

In the UK, the primary standard for absorbed dose for high-energy photon beams, such as those generated by linear accelerators, is a graphite calorimeter held by the National Physical Laboratory (NPL).

The calibration of field instruments is carried out first by calibration of a secondary-standard dosimeter at the NPL. The calibration of the secondary standard is in terms of absorbed dose to water and is carried out over a range of beam energies. For this purpose the index of beam energy is the quality index which is the ratio of dosimeter readings at depths of 10 and 20 cm for a field size of 10 cm × 10 cm and at a constant source to dosimeter distance. The calibration is then transferred by intercomparison of the field instrument and the secondary-standard dosimeter in the beam for which the field instrument will be used. In general the quality of the beam from a clinical machine will not be exactly the same as that at which the secondary standard was calibrated and it will be necessary to interpolate between (or extrapolate beyond) calibration factors which have been provided by the NPL. As long as the sensitivity of the secondary standard does not vary rapidly with beam quality this does not significantly increase the uncertainty on the calibration factors, which is ±1.5% at the 95% confidence level.

If access to a primary standard of absorbed dose is not possible the alternative is to make use of an exposure or air kerma standard based on a graphite walled, air filled ionization chamber. Such standards are available in many national standardizing laboratories, including the NPL, where dosimeters are calibrated by using a 2 MV x-ray beam. In most other countries a calibration in a cobalt-60 gamma-ray field is offered. The mean photon energies for these two radiation qualities are very similar.

The ionization chamber is fitted with a 5 mm thick plastic cap and calibrated in air against the national standard for radiation exposure. This yields a calibration factor $N$ such that

$$X = RN \qquad (12.1)$$

where $X$ is the exposure† and $R$ is the reading of the calibrated dosimeter corrected to 1013 mb and 20 °C. This correction is required if the ionization chamber is not air tight.

To utilize this calibration to measure absorbed dose at a point in a phantom the calibrated ionization chamber plus plastic cap is placed in the phantom as in figure 12.2(b). The dose at the centre position of the ionization chamber is then given by

$$D = RNC_\lambda \qquad (12.2)$$

where $D$ is the dose in centigrays, $R$ is the dosimeter reading corrected for temperature and pressure, $N$ is the calibration factor and $C_\lambda$ is a function of radiation quality. The factor $C_\lambda$ brings together the conversion from exposure to air kerma at 2 MV, the calibration energy and the variation of the sensitivity with energy of the secondary-standard ionization chamber for which $C_\lambda$ is calculated.

The derivation of $C_\lambda$ values and actual values is given in the HPA code of practice (1983). This document gives the uncertainty on absorbed dose measurements as ±2.3% for 2 MV x-rays to ±3.3% for 30 MV x-rays. This code also recommends that the ionization chamber be centred at a depth beyond the maximum of the depth dose curve. The dose at any point in the field may then be derived from isodose and depth dose data. The fundamentals of radiation dosimetry are discussed by Greening (1985).

† The SI unit of exposure is defined as ionization per unit mass in air, 1 C kg$^{-1}$. The older unit, the roentgen, is $2.58 \times 10^{-4}$ C kg$^{-1}$.

### 12.4.2. *Electron beam calibration*

For electron beam therapy, the calibration procedure is still basically the same as in figure 12.2(b) but further consideration has to be given to the type of ionization chamber employed and the procedures for electron beam calibration standardization are more complicated.

There is no calorimeter based absorbed dose standard as is the case for high-energy x-rays. The IPEMB code of practice for electron dosimetry (1996) is therefore based on an air kerma calibration. A series of additional factors is then applied to take into account the energy of the electrons at the point of measurement, the geometry and materials from which the chamber is constructed, the phantom material (if this is not water), polarity and ion recombination effects.

As with x-rays, dose per monitor unit for an electron beam is again a function of field size, and where the electron field is defined by fixed applicators each applicator will require a calibration factor to relate its dose per monitor unit to that for the reference field size.

### 12.4.3. *Calibration of the dose monitor for a multipurpose linear accelerator*

For a multipurpose machine, the calibration procedures will need to be carried out for each of the x-ray qualities and electron beam energies used. It is convenient to be able to adjust the sensitivity of the dose monitor to read 1 cGy per monitor unit for each radiation, at reference field size and SSD, as discussed in chapter 7.

## 12.5. SPECIFICATION OF DOSE RATE

Within very wide limits, dose rate is not critical to the success of a treatment carried out by radiotherapy, so specification of dose rate is determined mainly by operational convenience. The technology imposes no significant limits.

The total time to treat a patient is determined mainly by the time required to move the patient into and out of the treatment room and to set him or her up for treatment on the machine. This is likely to be something like 5–10 min, depending on the complexity of the particular set-up to be used. The typical dose for an individual treatment on a fractionated course is a few hundred centigrays. A further consideration is that a comfortably placed patient has to keep still during the treatment. All these factors lead to a useful exposure time per field of about a minute. Doubling or halving this time will not make a serious difference to the comfort of the patient during the treatment, or to the number of patients who can be treated during a working

**Table 12.1.** *Dosimetry protocols and codes of practice.*

| | | | Photons | Electrons |
|---|---|---|---|---|
| UK HPA[a] | 1983 | Revised code of practice for the dosimetry of 2–35 MV x-ray and of caesium-137 and cobalt-60 gamma-ray beams | ✓ | |
| UK IPSM[a] | 1990 | Code of practice for high energy photon therapy dosimetry based on the NPL absorbed dose calibration service | ✓ | |
| UK IPEMB[a] | 1996 | IPEMB code of practice for electron dosimetry... | | ✓ |
| IAEA | 1987 | Absorbed dose determination in photon and electron beams *Technical Report* 277 | ✓ | ✓ |
| USA AAPM | 1983 | A protocol for the determination of absorbed dose from high energy photon and electron beams | ✓ | ✓ |
| USA AAPM | 1994 | An extension of the 1983 AAPM protocol | | ✓ |
| Netherlands NCS | 1986 | Code of practice for the dosimetry of high energy photon beams *Report* NCS-2 | ✓ | |
| Netherlands NCS | 1989 | Code of practice for the dosimetry of high energy electron beams *Report* NCS-5 | | ✓ |
| Scandinavia NACP | 1980 | Procedures in external beam radiation therapy dosimetry with photon and electron beams... | ✓ | ✓ |
| Scandinavia NACP | 1981 | Electron beams with mean energies at the phantom surface below 15 MeV | | ✓ |
| Spain SEFM | 1987 | *SEFM Report* 2 | ✓ | |
| Italy AIFB | 1988 | Protocollo per la dosimetria... | ✓ | ✓ |
| France CFMRI | 1987 | Recommendations pour la mesure de la dose absorbée en radiotherapie... | ✓ | ✓ |

[a] HPA, IPSM and IPEMB are the same organization, which changed its name over the years covered by this table.

day. It follows that useful dose rates are in the range 200–500 cGy min$^{-1}$ at the isocentre.

For a particular machine it is useful to operate at a reasonably constant dose rate, perhaps within ±10%. This is desirable for two reasons:

(i) since a back-up timer is used in the dose monitoring system (see chapter 7), it is desirable to predict exposure times within this broad limit;

*Use of a radiation beam to demonstrate the position of the isocentre* 191

(ii) when the machine is being used as an x-ray generator, dose rate variations outside these limits are a likely indication of a malfunction. As previously discussed, when the machine is being used as an electron beam generator, a malfunction can conceivably increase the dose rate by a dangerous amount, so dose rate regulation is essential (see chapter 5).

*12.5.1. Dose rate and arc therapy*

When the gantry is rotated for arc or rotation therapy, the dose delivered per unit angle of rotation has to be constant, so that the prescribed dose is delivered while the gantry rotates between the stated angular limits. The rotation speed control setting is conveniently calibrated in dose monitor units per unit angle. This requires either that the dose rate be stabilized to meet the specified precision for dose monitor readings, which is a more exacting specification than the one just discussed, or that the speed of rotation of the gantry be regulated by a signal from the dose rate monitor, so that the monitor units per unit angle are constant to $\pm 2\%$.

## 12.6. USE OF A RADIATION BEAM TO DEMONSTRATE THE POSITION OF THE ISOCENTRE

The position of the isocentre and its tolerance is determined by the mechanical considerations discussed in the first part of this chapter. The ultimate object is to ensure that the central axis of the radiation field passes through the isocentre and this can be checked by the use of Kodak X-omat film. This film is chosen because it is very insensitive to x-rays and it requires an exposure corresponding to an absorbed dose of about 100 cGy to give controlled blackening on development. This is a realistic exposure for any test on a radiotherapy machine.

The film is exposed without intensifying screens, in a light tight envelope, and the position corresponding to the mechanical isocentre can be marked on the envelope in pencil. This will produce a visible pressure mark on the developed film. The film is set up in the plane of rotation of the radiation source with the pencil mark on the isocentre. This can be done with respect to the front pointer. One pair of collimators in the x-ray head is closed down to its minimum setting, or 2 mm, if this is possible, these collimators being set parallel to the axis of rotation of the gantry. The other collimators are set to an opening of several centimetres. A series of exposures are made at say 75° intervals of the gantry position, the gantry being static during the exposures. A satisfactory result will yield a developed film as in figure 12.7. This exercise should be repeated where the narrow beam recorded on the film is defined by the other pair of jaws. These tests should place the isocentre within the tolerance already mentioned, i.e. within a diameter of 4 mm.

**Figure 12.7.** *A radiograph showing narrow x-ray beams. Exposures are made at gantry angles which are multiples of 72°. This demonstrates that the central axis of the x-ray beam always passes through the isocentre.*

## 12.7. USE OF FILM TO SHOW THE POSITION AND SIZE OF THE X-RAY FIELD

A slow x-ray film in a light tight envelope is placed in the plane normal to the central axis of the radiation field through the isocentre. The envelope is covered with a Perspex sheet, into which lead markers have been inlaid, to define the corners of a $10 \times 10$ cm$^2$ field. The Perspex sheet should be equal in thickness to the build-up depth (the depth for maximum dose on the depth dose curve) for the x-ray quality concerned. The lead markers are lined up with the corners of the optical field of this size, and the film exposed and developed.

Measurements of the blackened area with a rule will give a measure of the field size to an accuracy of a few millimetres. Since they have been lined up with the optical field, the shadows of the lead markers and the edges of the blackened area will show the relationship between the positions of the x-ray and the optical fields to the same accuracy.

With adequately controlled exposure and development, the dose distribution across the radiation field can be determined in more detail by making densitometer measurements on the film and this will give results comparable to those in figure 8.1, which were measured with an ionization chamber.

## 12.8. GENERAL COMMENT

The specifications and performances discussed here refer only to the mechanical and radiation performance of the system as a whole. The individual circuits and units which make up the system are subject to their own detailed specifications, but these can only be discussed in relation to a particular machine and so are outside the scope of this chapter. Where the mechanical positions of the systems are read out via electrical transducers at the control positions, these need to be checked against direct mechanical measurements, but again the detail could only be considered in relation to a specific machine.

# CHAPTER 13

# RADIATION PROTECTION AND ROOM DESIGN

This chapter deals with two topics: protection of the patient from unwanted radiation inside the treatment room and protection of staff and the general public, where the potential hazard can arise both inside and outside the treatment room.

Protection of the patient requires consideration of machine design whereas protection of other personnel depends mainly on the design of the treatment room.

Although the design of a treatment room and associated facilities for a linear accelerator is normally the responsibility of an architect, he will need to be advised about the provision of radiation protection and about the usage of the equipment. Since this advice will usually be given by the physicist responsible for the linear accelerator, a section on this subject is justified. Some other issues concerning the design of the treatment room are also included in this chapter.

The legal authority in the UK, on which advice should be based, is given by the *Ionising Radiation Regulations* (1985). The *Approved Code of Practice for the Protection of Persons against Radiation arising from any Work Activity* (Health and Safety Commission) (1985) and the 'Guidance notes for the protection of persons against ionising radiations arising from medical and dental use' (NRPB 1988) are additional sources of information. However the specification for safety of medical electron accelerators is set out independently of any national legislation in IEC 601-2-1 (1981).

## 13.1. UNWANTED RADIATION

### 13.1.1. Leakage

Although x-ray emission for megavoltage radiation is primarily in the direction of the incoming electrons, there is a limited emission of radiation in other directions. The level of this radiation has to be controlled, partly to protect the patient from unwanted radiation outside the useful field and partly to limit the amount of shielding required from the walls of the treatment room. The specification of the amount of unwanted radiation is normally given in terms of leakage radiation, that is, the radiation measured when the radiation beam through the beam defining system is blocked off, either by closing the collimators completely or by closing them to minimum field size and then blocking off the beam with a sufficient thickness of lead. As far as the patient is concerned the leakage to be considered is that in a 2 m radius circular plane centred on, and orthogonal to, the beam axis, 1 m from the target. It is in this plane that the patient will be lying during treatment.

Leakage radiation is measured in air, using a detector with an appropriate build-up cap, and is expressed as a percentage of the dose rate at the isocentre of an unattenuated beam. From the description of the collimation system given in chapter 6 it is clear that the levels of leakage radiation will vary widely. There are two regions to be considered. IEC 601-2-1 (1981) specifies that the average leakage radiation in the region which is shielded only by the movable collimators (beam limiting devices in IEC terminology) should be less than 0.5% and the maximum leakage should be less than 2%. There is a relaxation for multi-element beam limiting devices, such as an MLC, such that the part of this region shielded only by the MLC should have an average leakage of less than 2% and a maximum leakage of less than 5%.

The second region comprises a surface, surrounding the accelerator and flight tubes, which is 1 m from the electron path, part of which is shown by the line EDC in figure 13.1. In considering radiation over this region, it should be remembered that any part of the structure which may be struck by an accelerated electron has to be regarded as a possible x-ray source. The whole accelerating waveguide, the beam guidance system and the target (and its surroundings) are therefore all possible sources of low-level radiation. IEC 601-2-1 (1981) specifies that the average leakage radiation in this region should not exceed 0.1% of the radiation level on the central axis of the beam and that the maximum leakage should not exceed 0.2%.

The waveguide and its surrounding structures, including the focusing coils, the deflecting magnets and the weight supporting structure of the gantry, provide some protection from this unwanted radiation. However it is still necessary to place additional shielding around all the possible radiation sources. Most of the shielding is lead, but where space is restricted tungsten is sometimes used as it has a density of 19 g cm$^{-3}$, which is much higher than 11.4 g cm$^{-3}$, the density of lead.

On machines which operate in more than one mode, i.e. at several x-ray and electron energies, it is necessary to consider leakage for each mode, although the worst case will usually be for the highest x-ray energy available.

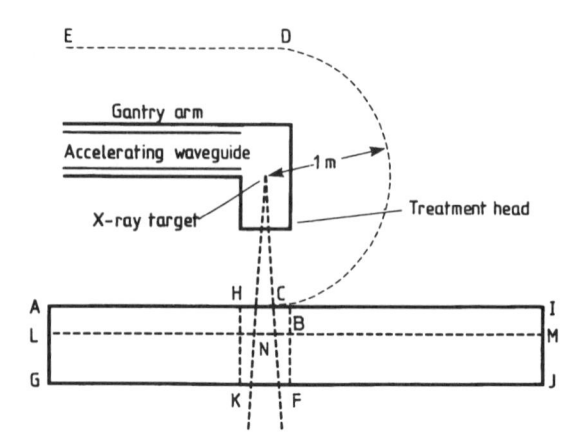

**Figure 13.1.** *Arrangements for measuring radiation levels outside the main beam of a linear accelerator. ABFG and HIJK represent alternative positions for a water phantom comparable in size to a patient's trunk.*

The dose levels to parts of the patient which are outside the target volume include contributions both from the leakage radiation, just discussed, and from scattered radiation from the beam defining system and the irradiated tissues inside the patient. Although it could be argued that these scattered components of unwanted irradiation are unavoidable it is discussed here so that leakage radiation can be seen in the context of total dose to the patient. The radiation distribution in a patient can be examined by placing a water phantom in position ABFG or HIJK in figure 13.1, where the phantom is comparable in length with the trunk of a patient. The dose distribution in a patient is represented by the measured distribution along the line LM, which is 10 cm deep in the phantom. It is also of interest to look at the dose distribution along this line in air arising from leakage and head scatter but obviously excluding scatter from within the patient.

Figure 13.2 shows the results of some measurements taken along the line NL of figure 13.1, for a $20 \times 20$ cm$^2$ field from a machine producing 8 MV x-rays. All the dose measurements are expressed as a percentage of the dose on the central axis and are plotted on a log scale as this shows up the levels outside the main beam. The leakage measurements (full circles) were made in air with the linear accelerator operating as an x-ray machine and with the main beam blocked off. This machine met the required specification as these leakage measurements were all below 0.2%. It can be seen that the dose levels in air (crosses) outside the nominal beam edge at 10 cm from the central axis are very high compared to the leakage levels in air, and even higher in the phantom (open circles) and that these levels exceed the leakage radiation up to a distance of 50 cm from the central axis of the beam.

Measurements of this type are discussed in more detail by Greene *et al*

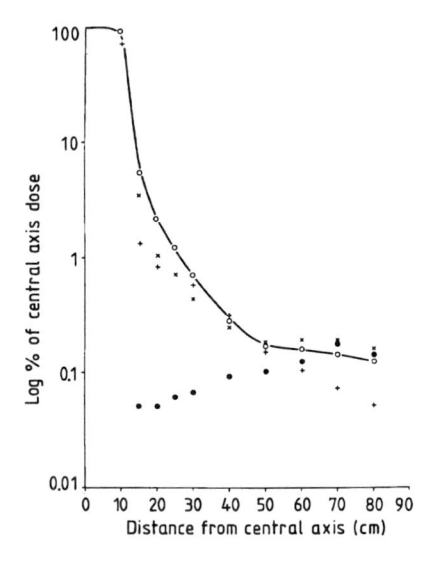

**Figure 13.2.** *Relative dose levels along the line NL of figure 13.1. Full circles, leakage radiation with the x-ray beam blocked off. Crosses, radiation levels in air for a 20 cm × 20 cm 8 MV x-ray beam. Open circles, radiation levels in a phantom for a 20 cm × 20 cm 8 MV x-ray beam. Plus signs, radiation levels in air for a 20 cm × 20 cm 14 MeV electron beam.*

(1983), who show that dose levels to a distance of 30 cm from the beam edge are determined mainly by scatter from the beam defining collimators at all field sizes, rather than by leakage radiation.

Measurements in the direction NM (figure 13.1) are comparable to those in figure 13.2 out to the 50 cm point but then continue to fall to levels approximately half of those in the NL direction. The dose levels in the NL direction are higher because of unwanted radiation from the accelerating waveguide. It is important to note that the dose due to scattered radiation varies with field size and that the dose gradient outside the useful beam is very steep.

Figure 13.2 also shows measurements made for a $20 \times 20$ cm$^2$ field of 14 MeV electrons from the same linear accelerator (plus signs) along the line NL in air. It can be seen that the dose levels outside the edge of the electron field are lower than those for the x-ray field. This arises because the much lower electron beam current in the accelerating waveguide required for electron therapy reduces the production of unwanted x-rays from the waveguide walls.

## 13.1.2. Neutrons and induced radioactivity

The reactions where a gamma-ray ejects a neutron from a nucleus ($\gamma$, $n$ reactions) have threshold photon energies for most elements in the region of

10–20 MeV, the maximum cross section for the reaction usually occurring at a photon energy of a few megaelectron volts above the threshold. For any particular reaction the neutrons have a continuous energy spectrum comparable to a fission spectrum and a mean neutron energy of less than 1 MeV (McCall 1981). The active product from a ($\gamma$, $n$) reaction usually decays by positron emission. The half-lives produced are usually short, of the order of 10 min. Electron–neutron (e, n) reactions may also occur but these have much smaller reaction cross sections than ($\gamma$, $n$) reactions. ($\gamma$, $p$) reactions have comparable threshold energies and cross sections to the ($\gamma$, $n$) reactions.

However the relatively low-energy protons produced are absorbed locally and, as the reaction products are stable, no radiation hazards arise.

Linear accelerators operating at electron energies above 10 MeV will produce neutrons and radioactivity by these processes. For a machine operating as an x-ray generator, the neutron output will be produced mainly by materials at or near the x-ray target where the photon fluence is maximum. The reaction cross sections are such that the neutron dose rate in the treatment field is of the order of 0.1% of the x-ray dose rate and can be regarded as a negligible component of the dose delivered to a patient. For references see the *British Journal of Radiology* (1983). The ($\gamma$, $n$) reactions will also produce measurable radioactivity in irradiated patients but the resultant dose to the patient will be small compared to the 0.1% component mentioned above.

The resultant activation of the machine components, although easily measured, is not a significant hazard to patients or staff when the machine is being set up for radiotherapy. However, if it is necessary to dismantle the treatment head for maintenance, components near the x-ray source, e.g. the x-ray target itself or the field flattening filter, may be sufficiently radioactive to require careful handling. Because of the short half-lives involved, this problem can be avoided by waiting for a few hours after the last irradiation. Also, because of the short half-lives, no long-term build-up of radioactivity is likely to occur.

Oxygen and nitrogen in the air of the treatment room will be activated by the ($\gamma$, $n$) reaction, but the resultant levels of radioactivity will be negligible in relation to maximum permissible values (Holloway and Cormack 1980).

Neutron production will also contribute to the unwanted radiation for a patient outside the treatment beam (Rogers and Van Dyke 1981). Although this only makes a marginal addition to x-ray dose delivered outside the treatment field (ICRP 1983) the maximum absorbed dose due to neutrons is specified in IEC 601-2-1 (1981) as 0.05%. At this level the overall risk to patients from neutrons inside and outside the treatment field is discussed by Epp *et al* (1984), who conclude that it is negligible compared with the risk of malignancies for the general population.

Although neutrons are not important in terms of dose to the patient they do warrant further consideration when designing treatment rooms.

## 13.2. THE TREATMENT ROOM AND ITS ENVIRONMENT

### 13.2.1. Treatment room layout

Figure 13.3(a) shows a plan view of the linear accelerator with its head in the two possible positions to give horizontal beams, with the edges of the maximum-sized x-ray fields shown by broken lines. The sections of the treatment room wall which intercept these maximum fields are called primary barriers and have to provide enough attenuation to protect personnel outside these walls. These personnel might be staff, members of the general public or other patients. For a machine which can rotate through 360°, primary barriers must be provided for all positions in the walls, floor and roof at which the beam may be directed (figure 13.3(b)), if it is possible for a person to be in line with the primary beam. If the roof is inaccessible the primary barrier in this direction may be of reduced thickness and clearly if the floor is built on solid earth there is no need to consider it any further.

The remainder of the walls, floor and roof have to provide protection from radiation from two sources: leakage radiation as described above, and scattered radiation. These sections of the structure are referred to as secondary barriers.

Leakage radiation from the treatment head has to be regarded as of the same quality as the primary beam. It will, of course, be subject to considerable filtration by the shielding built into the machine, but will also include scattered radiation from the shielding, and its detailed spectrum is usually unknown.

The major sources of scattered radiation are the patient who is being irradiated and the portions of the walls, floor or roof which intercept the primary beam. This radiation is of lower mean photon energy than the primary.

The radiation incident on the secondary barriers is then a mixture produced by the very complex processes just outlined and originating from rather diffuse sources. It would be very difficult to determine its spectrum either theoretically or experimentally. Since the leakage radiation is of higher mean photon energy than the scattered component, and is of comparable magnitude, a secondary barrier which is adequate to attenuate the leakage radiation will also be able to deal with the scattered radiation component. Experience shows that this statement is an adequate basis for secondary-barrier design.

The dose rate of the leakage radiation at the secondary barriers is determined by the specification of the machine. This states that the leakage

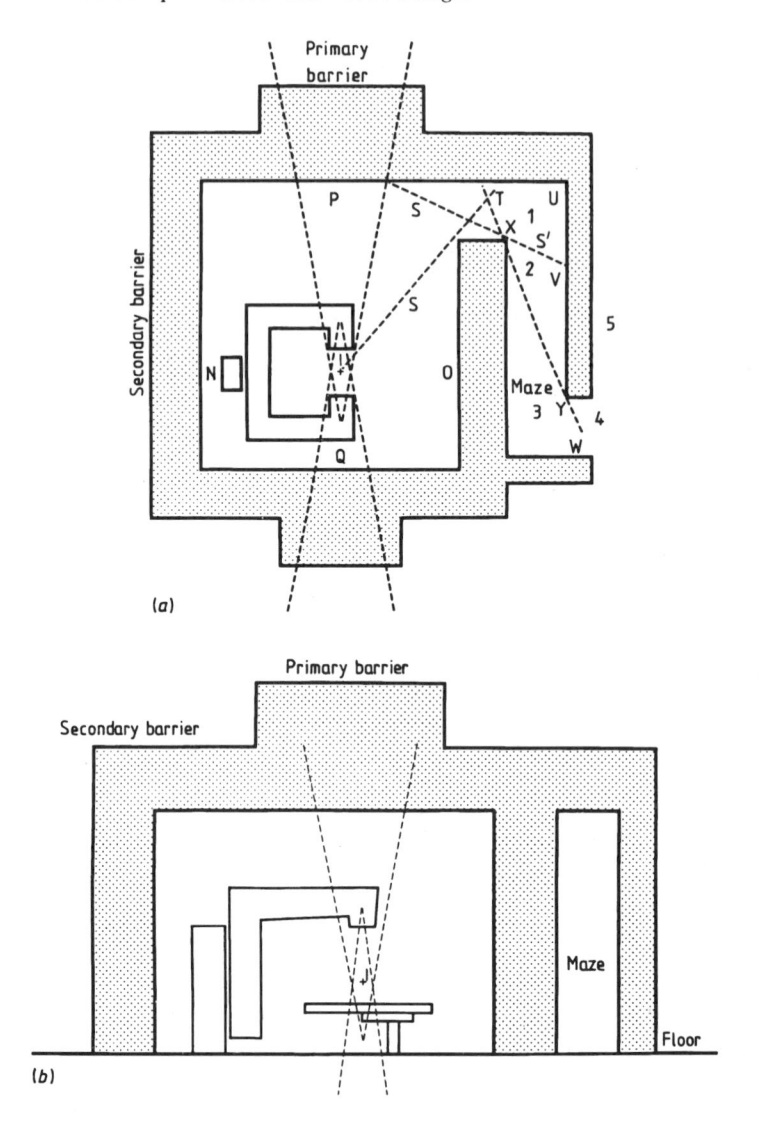

**Figure 13.3.** (a) A plan view of a treatment room, showing a linear accelerator with its head in two possible horizontal beam positions. The dotted lines from the head represent the edges of the largest beam available. (b) A vertical section of a treatment room through the isocentre in the plane NO of (a).

dose rate at 1 m from the target shall be not greater than 0.2% of that in the main beam at the same distance from the target. The leakage radiation dose rate at the inside of the barriers is approximately related to that at 1 m from the target by the inverse square law. This relationship is only approximate because the source of radiation is diffuse, being the whole shielding structure of the linear accelerator.

This section can be summarized as follows.

(i) The primary barrier thickness is chosen to attenuate the highest-energy x-rays at the maximum dose rate that the machine can produce, at the position of the barrier. This dose rate is related to the specified dose rate from the machine, usually stated at 1 m from the x-ray target, by the inverse square law.
(ii) The secondary-barrier thickness is chosen to attenuate the specified leakage x-rays from the machine when it is operating at maximum dose rate.

This discussion so far has been about a linear accelerator to be used as an x-ray generator. For a dual purpose machine three further comments are required.

(i) In the electron beam mode, the x-ray primary barriers will be more than adequate to stop the primary electron beam, which has a limited range.
(ii) A similar comment can be made about scattered electrons arriving at the secondary barrier.
(iii) In the electron therapy mode the electron current in the accelerating waveguide is down by a factor of a hundred on that required in the x-ray mode. Any unwanted x-ray production will then be less than that in the x-ray mode and therefore more than adequately attenuated by the primary and secondary barriers required in the x-ray mode.

*13.2.2. Access to the treatment room*

It is clear that access to the treatment room should be through a secondary barrier. The two possible systems are a suitable door or an entrance maze.

A door which will supply the same level of radiation protection as the secondary barrier for high-energy x-rays will weigh many tonnes and will have to be a ponderous, power driven, device. In a radiotherapy department this door will have to be operated something like a hundred times each working day, so its operating time is likely to be a problem. Further, the possibility of a mechanically driven door jamming with a patient inside the room is rather alarming. For these reasons shielded doors are rarely, if ever, used.

Figure 13.3(a) shows the plan view of an entrance maze. The inner wall forming the maze has to be comparable in thickness to the other secondary barriers, because this forms the total shielding for a person immediately outside the maze entrance (position 4).

The only radiation which can directly enter maze is scattered radiation either from the patient or from the irradiated part of the primary barrier. The worst situation that can arise is that shown in figure 13.3(a) where the

**Table 13.1.** *Measured dose rates at positions marked on figure 13.3(a).*

| Positions | Measured dose rate ($\mu$ Gy h$^{-1}$) |
|---|---|
| 1 | 3400 |
| 2 | 500 |
| 3 | 80 |
| 4 | 4.5 |
| 5 | 0.9 |

beam is directed at the primary barrier marked P, i.e. the one nearest to the maze. Scattered radiation can then enter maze from the patient or the primary barrier, as indicated by the lines marked S. The line SS' limits the region in the maze which can receive 'first-scatter radiation' and the wall section TUV shows the region which can scatter this radiation again along the maze. An empirical rule for designing the rest of the maze is that it must not be possible to see any of this wall section from outside. The length of the maze in the direction VW is then determined by the line XY. It can be seen in figure 13.3(a) that XY is not quite the optical line of sight, but takes account of the fact that the scattered radiation can penetrate the corners of the walls forming the maze. XY has been chosen so that the limiting scattered ray has to penetrate about 20 cm of concrete at each end. The width of the maze corridor, and its entrance and exit, are determined by the need to be able to wheel a patient on a standard hospital trolley through it. The outer wall of the maze, VY, is required to attenuate scattered radiation and does not need to be as thick as the inner wall. Experience suggests that it needs to be about half the thickness of the inner wall.

The principles outlined for the design of the maze can be justified by quoting some measurements made in a room very similar to that shown in figure 13.3(a). The room housed a 4 MV linear accelerator and the measurements were made with a maximum-size field directed at the primary barrier P. The machine output at the isocentre was x-rays at 500 cGy min$^{-1}$.

Table 13.1 shows measured dose rates at the positions marked on figure 13.3(a). Position 2 was 0.5 m outside the line SS', where there is a sharp drop in dose rate, and position 4 was 0.5 m outside the line XY, where again the dose rate drops rapidly. The 'quality factor' for x-rays is unity so these measurements are numerically equal to the equivalent dose in microsieverts[†] per hour. The dose rate at the control unit (see figure 13.4), the position

---

[†] Equivalent dose is defined as dose in grays multiplied by a quality factor. The quality factor takes account of the different biological effects produced by unit doses of different radiations and values are given by the ICRP (1991). For x-rays the quality factor is unity. The unit of equivalent dose is the sievert (Sv).

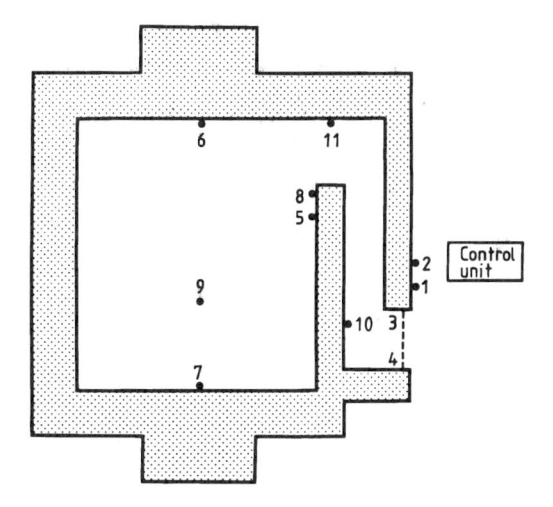

**Figure 13.4.** A plan view of a treatment room. The numbered points are described in the text.

occupied by staff when irradiations are being performed, is 0.9 $\mu$Gy h$^{-1}$. The mean dose rates during a working day are about a factor of ten down on these values because the machine is only generating radiation for about 25% of the time, and also because the dose rates are much lower when the primary beam is not directed at the primary barrier P.

### 13.2.3. The dimensions of the treatment room

Referring again to figure 13.3(a), the machine can be placed fairly close to the position N, allowing 30–50 cm between the back of the support pedestal and the wall as access for servicing. The dimensions IO, where I is the isocentre, have to exceed the length of a patient who may have his head at I and his feet pointing towards O, and allow enough space for staff to pass between the end of the treatment couch and the wall. Consequently the distance IO needs to be a minimum of 2.5 m. These considerations fix the dimension NO. It is usually advantageous to place the machine off centre in the direction PQ, placing it so that the gantry clears the wall at Q when it is in the horizontal beam position and again allows about 30–50 cm for access by service personnel. The distance IP may then be arrived at by considering the largest field size that is likely to be required. If a field for whole-body treatment is required for a machine with a maximum field size at 1 m SSD of $40 \times 40$ cm$^2$, then IP needs to be about 5 m to allow a recumbent patient to be treated. These considerations fix the dimensions PQ of the treatment room to about 7 m. There are, of course, no technical reasons why the room should not be larger than the sizes discussed, but normally both space and cost considerations will limit the room size to the minimum which is required for utilization of the treatment machine.

The height of the treatment room is determined mainly by the height of the gantry in the upright position (see figure 13.3(b)). However, for installation and servicing the equipment it is useful to install a lifting beam which passes over the isocentre in the direction NO, and which can carry a block and tackle or other lifting gear. This will add to the height of the underside of the protected roof, which will be typically 3 m.

### 13.2.4. Detailed design of protective structures

The material used for the protective walls of the treatment room and maze is determined by building, space and cost considerations, and is normally ordinary concrete. Where there is insufficient space available for the wall thickness that would be required for ordinary concrete, it may be necessary to use a dense concrete, for example concrete based on a barium containing aggregate. In either case it will be necessary to specify the density of the dried-out concrete, as well as the wall thickness required. It would, of course, be possible to reduce the thickness of protective barriers still further by using higher-density materials, e.g. steel or lead, but this is usually not feasible for structural and cost reasons. It is also possible to use soil as some of the protective material required by placing the treatment room underground, although it usually appears to be more economical to use concrete rather than make excavations purely to provide radiation shielding.

### 13.2.5. Shielding data and determination of barrier thickness

For shielding calculations, broad-beam attenuation data are required. The dose rate outside a radiation absorber increases with field size due to an increasing contribution from radiation scattered in the absorbing material. At very large field sizes the increase in scatter with field size is balanced by the increasing obliquity, and therefore absorption of the scattered radiation, so that the transmitted dose rate becomes more or less independent of field size. For practical purposes the radiation can be considered as being attenuated exponentially in the shielding materials and for shielding calculations it is convenient to express the attenuation data in the form of a 'tenth-value thickness' (TVT), the thickness of material required to reduce the transmitted dose rate by a factor of ten.

Table 13.2 gives TVT as a function of nominal accelerating voltage (MV), for x-rays in ordinary concrete (density 2350 kg m$^{-3}$), barytes concrete (density 3500 kg m$^{-3}$), steel and lead.

Published data for a wide range of materials are not readily available but, to a reasonable approximation, the TVT is reduced from that for normal concrete by the ratio of the densities. This is a safe approximation,

**Table 13.2.** *TVT (cm) for shielding materials as a function of x-ray energy (nominal accelerating voltage).*

| Energy (MV) | Standard concrete 2350 kg m$^{-3}$ | Barytes concrete 3500 kg m$^{-3}$ | Steel 7800 kg m$^{-3}$ | Lead 11 400 kg m$^{-3}$ |
|---|---|---|---|---|
| 4 | 29 | 20 | 9 | 5 |
| 6 | 34 | 23 | 10 | 5.5 |
| 8 | 36 | 24 | 10 | 5.5 |
| 10 | 38 | 25.5 | 11 | 5 |
| 16 | 42 | 28 | 11 | 5 |
| 25 | 46 | 31 | 11 | 4 |

which, if anything, will over-estimate the TVT because it ignores the additional attenuation due to photoelectric absorption in higher-atomic-number material.

The factors which determine the maximum allowable instantaneous dose rates in areas outside the barriers are the following.

(i) *Access to these areas.* Are they controlled?
(ii) *Occupancy of these areas.* Are those individuals who have access to these spaces members of staff† or the general public? The dose limits for the general public being lower than those for occupationally exposed workers.
(iii) *Duty cycle and orientation of the accelerator.* For what fraction of the working day is any particular part of the barrier actually irradiated? Note that time averaging is only acceptable in areas where the instantaneous dose rate is less than 2 mSv h$^{-1}$ and then the time averaging should be over a period not exceeding 8 h.

Having determined the maximum allowable instantaneous dose rates the shielding can be determined from the following additional factors:

(i) the dose rate at the inside of the barrier when it is being irradiated;
(ii) the maximum energy of the radiation incident on the barrier;
(iii) the attenuating properties of the shielding material.

Statutory dose limits, set out in national legislation, are based on the quantitative estimates of the consequences of radiation exposures and recommendations of the International Commission on Radiological Protection (ICRP). The most recent recommendations and rationale are

---

† In principle staff can be separated into two groups, classified and unclassified workers, where those who are classified are likely to receive doses exceeding three-tenths of the annual dose limit. In practice staff working with linear accelerators are rarely classified as there should never be a need for them to occupy spaces where the radiation exposure would lead to such doses.

given in *ICRP Publication* 60 (1991). The dose limit for occupationally exposed workers should be 20 mSv/year, averaged over 5 years, with a further provision that the dose should not exceed 50 mSv (the earlier limit) in a single year. The dose limit for members of the public should be 1 mSv/year if they are likely to be continuously exposed and should not exceed 5 mSv in a single year if exposure is infrequent. Over and above limitations based on dose, the earlier recommendations in *ICRP Publication* 26 (1977) suggest that dose rates to radiation workers and the public be made 'as low as reasonably achievable', this being referred to in the health physics literature as ALARA‡. In the present context, the ALARA principle suggests that concrete barrier thicknesses calculated in accordance with the factors mentioned should be regarded as minima, and that additional concrete should be used if possible. In practice the limitation on additional barrier thickness is likely to be space rather than cost. If the necessary decisions about thickening up barriers are made at the planning stage this will result in only marginal additional costs.

Utilization of the available information to estimate the thickness required for radiation barriers may be illustrated by example.

*Example 1.* Calculate the thickness of ordinary concrete required for a primary barrier whose outer face is 7.1 m (this distance is chosen for convenience of calculation: $7.1^2 \simeq 50$) from the isocentre of a 10 MV x-ray generator operating at a dose rate of 4 Gy min$^{-1}$ at 1 m from the target.

The unattenuated dose rate beyond the barrier, when the radiation beam is incident on it, is $(4 \times 60)/50 = 4.8$ Gy h$^{-1}$ (which for x-rays equates to 4.8 Sv h$^{-1}$). If the machine can be rotated through 360°, the radiation beam will only be incident on the portion of the primary barrier concerned for a fraction of the time, say 25%. Also, on the radiotherapy duty cycle, the machine will produce radiation for about two hours out of an eight-hour working day, the rest of the time being taken up in moving patients in and out of the treatment room and setting them up for treatment. The barrier will then be irradiated by the primary beam for half an hour per day.

If the area outside the barrier is occupied for the whole of the working day by staff who are not classified as radiation workers their annual dose should be much less than the dose limit (20 mSv), say 5 mSv.

Over 250 working days per year, each 8 h, 5 mSv per year would be accumulated from a continuous instantaneous dose rate of 2.5 $\mu$Sv h$^{-1}$.

Taking into account the duty factor of 25% and the beam orientation factor of 25% the maximum instantaneous dose rate allowed is 40 $\mu$Gy h$^{-1}$. (Good practice allows time averaging as long as the instantaneous dose rate does not exceed 2 mSv h$^{-1}$.)

‡ In UK legislation the word achievable has been replaced by practicable and ALARA becomes ALARP. This change recognises that additional shielding will always achieve a lower dose rate but that the cost–benefit ratio will become progressively higher.

The barrier has to reduce the dose rate by a factor of $4.8/(40 \times 10^{-6}) = 1.2 \times 10^5$.

The TVT for 10 MV x-rays in concrete of density 2350 kg m$^{-3}$ is 38 cm (table 13.2). The number of TVTs required is $\log_{10}(1.2 \times 10^5)$ which is 5.08. The thickness of the concrete barrier required is therefore $5.08 \times 0.38$ m = 1.93 m. It would be reasonable to round this up to 2 m.

To apply the ALARA principle this thickness could be increased by up to another TVT if space were available.

*Example 2.* Calculate the thickness of a secondary barrier whose outer face is at 5 m from the isocentre for the same 10 MV x-ray generator, where the leakage dose rate is 0.2% of that in the main beam at 1 m from the target.

The secondary barrier is exposed to leakage radiation on all occasions when the beam is switched on, so the beam orientation factor cannot be included. The maximum instantaneous dose rate outside this barrier is therefore one-quarter of that outside the primary barrier, if radiation workers are present all day, i.e. 10 $\mu$Gy h$^{-1}$.

The unattenuated dose rate from leakage radiation is $(4 \times 0.2 \times 10^{-2} \times 60)/50 = 9.6 \times$ mGy h$^{-1}$.

The barrier has to reduce the dose rate by a factor of $9.6 \times 10^2$. This can be achieved with three TVTs, which in this case is 1.14 m.

If the spaces considered are to be accessible to non-radiation workers, this thickness should be increased by another TVT to ensure that this group also receives a dose of much less than their dose limit.

All the above discussion assumes that the walls of the treatment room are made of cast concrete which meets the density specification. If the use of concrete blocks is more convenient the joints will have to be staggered and well mortared. Even if the blocks meet the density specification it is likely that the overall density will be reduced and the thickness of the walls will have to be increased by a small amount to allow for this.

In principle the design of the protective barriers for the floor and roof of the treatment room is dealt with in exactly the same way as that of the walls. If the treatment room is on ground floor level then solid earth underneath will require no protection. If, as is not unusual, the room is over a basement housing heating and ventilation equipment, this may not require protection provided the entrance to this space is interlocked to prevent the production of radiation when the basement is occupied.

If the treatment room is part of a single-storey building with adequately controlled roof access it will still be necessary to provide some roof protection to prevent 'air shine', i.e. to make sure that adjacent spaces are not significantly irradiated as a result of scatter, by air, from the radiation penetrating the roof.

It is, of course, necessary to have holes in the protective structure to bring in electrical and water supplies and ventilation. These holes should clearly not be in the primary barriers. The holes for electricity and water can be of relatively small section and do not present any serious problem, so long as they do not point directly at the radiation source. The holes for ventilation in and out are likely to be of larger section and their location, and the provision of the necessary protection round them, requires careful consideration.

The design of radiation protection is further discussed in *IPEM Report* 75 (1997) and in *NCRP Report* 49 (1976).

## 13.3. NEUTRON SHIELDING

Neutron production by linear accelerators generating photons of energy in excess of 10 MeV has been briefly discussed earlier in the context of unwanted irradiation of the patient. The general problems related to neutron production and protection are covered by *NCRP Report* 79 (1984). In respect of protection of staff and the general public it is generally the case that the concrete structure required for x-ray protection will provides adequate protection from neutron radiation. However further consideration has to be given to neutrons in respect of the design of the maze. As neutrons interact with the shielding materials they are slowed down and scatter around the corners of the maze. The maze is therefore less effective for neutrons than for x-rays so that although the contribution to dose from neutrons is negligible inside the treatment room it might be the dominant source of exposure at the maze entrance. An empirical method of calculating the attenuation of neutrons in treatment room mazes has been described by Kersey (1979). It can be assumed that the neutron dose rate falls according to the inverse square law from the isocentre to the inner end of the maze, point T in figure 13.3(a). From that point onwards the concept of a tenth-value length (TVL) is employed. This is analogous to a TVT: it is the length of maze required to reduce the dose rate from neutrons by a factor of ten. The dose rate at the outer end of the maze is also proportional to the minimum cross sectional area.

For a maze with a cross sectional area of 6 m$^2$ the first TVL is approximately 3 m and subsequent TVLs are 5 m (the first TVL is less because the neutrons have not slowed down very much at this point).

*Example*: A linear accelerator operates at 15 MV. The x-ray dose rate at the isocentre is 500 cGy min$^{-1}$ and the neutron component is 0.1% i.e. 0.5 cGy min$^{-1}$, 0.3 Gy h$^{-1}$.

The distance from the isocentre to the inner end of the maze is 5 m and the length of the maze is 8 m. The maze has a minimum cross section of 4.5 m$^2$.

The neutron dose rate at the outer end of the maze is given by

$$D_n = \frac{0.3 \times 4.5}{5^2 \times 10^2 \times 6} \text{ Gy h}^{-1} = 0.09 \text{ mGy h}^{-1}.$$

The radiation weighting factor (previously known as the quality factor) for neutrons is between five and twenty depending on the neutron spectrum (ICRP 1991). Making the safest assumption that it is twenty, the equivalent dose rate due to neutrons will be 1.8 mSv h$^{-1}$. At this level time averaging is not sufficient to reduce the equivalent dose rate to 10 $\mu$Sv h$^{-1}$, which is necessary for uncontrolled access. It is therefore necessary to consider additional means of reducing the dose rate. The alternatives are the following:

(i) lengthen the maze, but in this case a reduction by a factor of 180 requires an additional 2.25 TVLs, which be impractical;
(ii) reduce the minimum cross section with concrete lintels (impractical for such a large reduction);
(iii) line the maze with a material which will absorb neutrons such as boron load polythene or borated plaster or
(iv) close the maze with a 'neutron door'.

A neutron door can be constructed from borated polythene covered by lead or steel, the borated polythene serving to slow and capture the neutrons and the steel or lead to absorb gamma-rays generated from $(n, \gamma)$ reactions. Typically a neutron door consists of between 5 and 10 cm borated polythene covered by 1–2 cm lead or steel. The weight of such a door is very high, typically 600 kg, and to be practicable such doors have to be motorized. The failure of the mechanism is another mechanical hazard that has to be assessed in the design of a room for a high-energy accelerator.

## 13.4. MISCELLANEOUS POINTS ABOUT TREATMENT ROOM SERVICES

Current machine design places the modulator and most of the electrical and electronic circuits immediately adjacent to the gantry support structure, inside the treatment room. The control system needs to be outside the treatment room near to its entrance. The remaining unit, the water cooler, needs to be in an area where it can exchange heat to the atmosphere. Partly for this reason, and also because it is likely to incorporate noisy water pumps, it is usually placed outside the treatment room and away from the control area. Apart from these considerations the placing of the water cooling unit is not critical, there being no major limitations on the lengths of the water pipes and cables required to connect it to the linear accelerator.

### 13.4.1. Control unit

This has to be outside the treatment room and should be located near the maze entrance, for example as shown in figure 13.4(a), partly because the operators have to move frequently from the control position into the treatment room and also because the person controlling the machine should keep an eye on the entrance.

### 13.4.2. Control of access

It is necessary to have an interlock system which switches off the radiation if an unauthorized person enters the treatment room. This can be a door-operated switch at the maze entrance or a photocell interlock operated by a series of light beams at different heights at position 3–4 in figure 13.4. Light beam interlocks have to be very carefully designed to ensure that they fail safe, are continuously self-checking and cannot be affected by ambient light. If the maze is designed to reduce the dose from x-rays as outlined in this chapter the door serves no protective purpose so the photocell system will serve adequately and save the operators opening and closing the door many times a day. If a neutron door is required then the interlock will be two independent switches operated mechanically by the door. Best practice requires that the room entrance interlock includes a 'search and lock-up facility'. Setting the interlock requires two actions: first the operator has to enter the room to ensure that nobody, other than the patient, is inside. A push button switch at position 5 at the inner end of the maze is operated to start the sequence. At this time an alarm is sounded in the room to alert anyone who may not have been seen. The operator and any other personnel then have to leave the room within a pre-set time, typically 15 s and operate a second push button switch at postion 1 which closes the interlock circuit. Thereafter anyone entering the room will break the interlock and the complete sequence will have to be repeated.

If this system fails, either because of a mistake by the responsible person or because of a fault in the interlock system, someone other than the patient may be left in the treatment room when the radiation is switched on. The machine is required (IEC 1981) to produce a noise to indicate that radiation is being produced or is about to be produced. A linear accelerator will do this in any case, the noise frequency corresponding to the pulses from the modulator at a frequency of typically 200 Hz. The person accidentally left in the room can react to the noise by pressing the 'emergency off' switch discussed below.

### 13.4.3. Warning lights and signs

It is required that warning lights be switched on when the machine is producing radiation (IEC 1981): one inside the room and one adjacent to the entrance. These can be controlled from the circuit which energizes the modulator, and they are located at positions 10 and 11 in figure 13.4. It is also necessary to display the trefoil radiation warning sign adjacent to the maze entrance.

## 13.4.4. Power supplies

It is convenient to bring all the power supplies for the machine, except those for the vacuum system, through a single contactor. The vacuum system needs a separate supply because it is normally energized continuously. All the remaining systems need only be energized during the working day. In chapters 9 and 10 it has been suggested that the main contactor for these systems should be opened by one of a set of push buttons in an emergency. Positions 2 and 5 in figure 13.4 are suitable for two emergency push buttons for this purpose. Some treatment machines are set into a wall between an 'equipment space' and the treatment room. In this case emergency off switches have to be available in both spaces. Similarly if the patient support system is mounted in a pit which is large enough for access by maintenance personnel this area should also be fitted with emergency off switches.

## 13.4.5. Axis lights

It is useful to have fixed laser pointers converging on the isocentre provided by a pair of horizontal laser beams from positions 6 and 7 in figure 13.4 and one pointing vertically downwards from position 9. Because these beams behave as long optical levers, which magnify any movement of the source, they need to be very rigidly mounted. These lights are discussed in section 8.4.2.

## 13.4.6. Room lighting

To facilitate the use of the various optical pointers as well as the lights just mentioned, it is useful to have a two-level lighting system in the treatment room. It is convenient to have the switch for this system on the pedestal or pendant which carries the machine controls inside the treatment room.

## 13.4.7. Ventilation

No special ventilation problems arise from the use of a linear accelerator as a radiation source. Ozone production by the radiation beam is not a significant problem (Holloway and Cormack 1980). Ozone production due to malfunctions of the high-voltage components resulting in sparking will be controlled by the protective circuits built into the modulator (see chapter 3). They should shut down the HT supply under these conditions. The production of radioactivity in the air by high-energy photon beams has been mentioned earlier in this chapter and is again not a significant hazard to patients or staff.

## 13.4.8. Patient viewing and communication

The patient has to be alone in the treatment room during actual irradiations but needs to be kept under observation. This is conveniently and most economically done using one or more closed circuit TV systems with the monitors placed adjacent to the control unit. It may also be desirable to have an intercom system for communication with the patient.

### 13.4.9. Access for the machine

Provision has to be made for getting the components that make up the radiotherapy machine into the treatment room. The detailed way in which this is organized depends mainly on the size of the units into which the gantry is broken down for delivery and installation.

For a new treatment room, the simplest system is to have a suitable sized hole in a secondary barrier, and fill it up with concrete blocks after the machine has been placed in position. After a very long period of use this hole may be opened to replace the equipment. This presupposes that the secondary barrier is part of an outside wall of the building to which access is available for equipment delivery.

For a new room where access through a hole in a secondary barrier is not available it may be necessary to design the maze to allow passage of the largest component or, alternatively, to construct part or all of the inner maze wall (adjacent to point O in figure 13.3(a)) after the machine has been installed. Since the treatment room is likely to be used for more than one linear accelerator over a long period of time, it is probably useful to have this wall constructed of concrete blocks, to facilitate dismantling it for installation of a new machine at a later date.

## 13.5. GENERAL COMMENT

This chapter has considered the design of the treatment room to house a linear accelerator from the point of view of efficient operation and utilization of the machine, and in relation to the provision of radiation protection. The many other design factors affecting the comfort and convenience of patients and staff are outside the scope of this book

# CHAPTER 14

# ACCELERATOR OPERATION

This chapter concentrates on operational issues which underpin the successful use of a linear accelerator as a radiotherapy machine rather than dealing with its actual use for treating patients. These issues include commissioning, routine quality control measurements, planned preventive maintenance and operating costs.

## 14.1. COMMISSIONING

The procedures involved in testing the mechanical properties and the radiation output of a new linear accelerator have been outlined in chapter 12. These tests, to ensure that the machine meets the contractual specification, are strictly speaking acceptance tests. In addition to this commissioning includes collection and application of the large body of information which is required for treatment planning and for calculation of machine settings for actual treatments. This includes various output factors, depth dose data and isodose charts or the equivalent data stored in a treatment planning computer.

Output factors are those which relate the monitor dosimeter readings to patient dose. They may be stated in units of absorbed dose per monitor unit. The obvious example of such an output factor is the dose per monitor unit measured at a defined reference point in a phantom irradiated by the x-ray beam of an accelerator. Further examples are output factors for each electron applicator at each energy.

Other output factors are relative: these include wedge attenuation factors and field area factors, which are dimensionless factors giving the ratios of two outputs. In these examples the wedge attenuation factor gives the ratio of dose per monitor unit with and without each wedge filter in position and the field size factor gives the ratio of dose per monitor unit for each field size relative to a reference field size, say 10 cm × 10 cm.

Percentage depth dose tables are a special case of relative output factors, which relate the dose as a function of depth to the maximum dose at each field area. Comprehensive reviews of percentage depth dose for a wide range of energy and for different types of machine have been carried out, particularly by the British Institute of Radiology. Compilations of depth dose data have been published, most recently in supplement 25 of the *British Journal of Radiology* (1996).

Although the published data are intended to be indicative rather than definitive there is a case for making a limited set of measurements to confirm the validity of the published data when a new machine is being commissioned. If the agreement is acceptable, over the full range of depths and field sizes to be used, then adoption of the published data will avoid the collection of a large number of data.

Similarly if the machine being commissioned is nominally identical to an existing machine in the department it is usually sufficient to make a limited number of depth dose measurements having carefully established the depth dose at 10 cm deep (or other index of energy) in order to validate the use of existing data for the new machine.

Measurement of this large collection of data is a major exercise and is clearly not required in detail for every new machine. It is of course, required if a new make or model of linear accelerator is being commissioned. Percentage depth dose tables are necessary for fixed-SSD treatments; the related tables of tissue–phantom ratios (TPRs) or tissue–maximum ratios (TMRs) are needed for isocentric treatments and can be measured directly or calculated from percentage depth dose tables and peak scatter factors.

The data required to produce a full description of the radiation beams in two or three dimensions depend mainly on the beam model employed in the treatment planning computer systems that will be used.

The simplest systems require the mass storage of an array of data for each beam measured. In this case the dose distribution in a phantom can be reasonably accurately represented by the central axis percentage depth dose values and dose profiles measured normal to this axis at four or five depths including $d_{max}$. The dose calculation algorithm then provides for interpolation between field sizes and for calculation of the dose at any arbitrary point.

More complicated systems model the primary beam separately from scattered radiation. In these cases the volume of data is reduced to 'in air' dose profiles, zero-area tissue–air ratios (TARs) and differential scatter–air ratios (SARs).

Finally the most complicated systems model radiation beams by convolving the dose distribution for a pencil beam with functions describing the geometry and radiation scattering characteristics of the collimators. The data required for these systems are again limited and are not necessarily easily identified with measurements of dose distributions, the purpose of the data being to provide the constants required for the model to accurately reproduce actual dose distributions. A fuller discussion of the different types of algorithm that have been employed is given by Redpath et al (1993).

Even if the treatment planning model does not require a large number of input data, it will usually be necessary to collect data in a form similar to that described for the simple system so that the output dose distribution generated by the treatment planning system can be checked against actual dose distributions measured in a phantom. For this purpose the traditional representation of dose distributions as a family of two-dimensional isodose charts produced in the form of hard copies is invaluable.

Three-dimensional computerized beam data acquisition systems are available to carry out this task and are often pre-programmed to collect data in the format required for particular treatment planning systems. Transfer of these data from the beam data acquisition system to the planning system can be via network connection or floppy disk.

Although these commissioning measurements have been described in relation to x-ray beams a similar number of data will need to be collected for electron beams for each applicator and beam energy.

## 14.2. ROUTINE RADIATION MEASUREMENTS

### 14.2.1. X-ray measurements

The success of the large team effort, by the radiotherapist, physicists and radiographers, which makes up a radiotherapy treatment depends finally on the delivery of the correct radiation dose. It has been shown clinically that a 5% change in dose will significantly change the outcome of a radiation treatment (Stewart and Jackson 1975). Thus routine checks on the factors which determine the doses delivered to patients have to ensure that these remain constant to much better than 5%, say 2%.

Radiation field measurements were discussed in the last chapter in terms of measurements in a water phantom and this is appropriate for commissioning and initial calibration of the dose monitor. However, the use of a large water phantom is not necessary for routine measurements, a solid phantom being much more convenient. This phantom does not need to be exactly water or tissue equivalent and should be of approximately unit density and of low-atomic-number material. Perspex (polymethylmethacrylate) is readily available but is somewhat denser than water; polystyrene is near unit density but is only readily available in thin sheets. It is also very expensive as are specially formulated tissue equivalent plastics. The material chosen can be machined into a 30 cm × 30 cm × 20 cm block (figure 14.1) or constructed in the form of a sealed tank which can be filled with water.

The top is conveniently painted white and marked with a centre and two concentric squares of 10 cm × 10 cm and 20 cm × 20 cm. This surface and

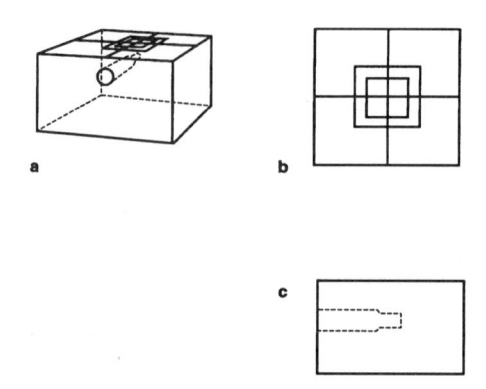

**Figure 14.1.** (a) A phantom for routine radiation measurements in x-ray beams. (b) A top view of (a). (c) A vertical section through (a).

markings provide a reference system for routine checking of the centring of the front pointer, and then the size and centring of the optical beam.

Figure 14.1(c) shows a vertical section through the routine check phantom, where a single hole is drilled to hold a suitable dosimeter such as a Farmer ionization chamber. The hole is drilled to centre the chamber under the centre mark on the surface (figure 14.1(b)) at the recommended depth of measurement for the radiation quality concerned, i.e. at 5 cm deep for qualities up to 10 MV, 7 cm for qualities 11–25 MV, and 10 cm for qualities above 26 MV.

The basic calibration of the dose monitor system is carried out with a National Physical Laboratory (NPL) calibrated secondary-standard dosimeter in a water phantom as discussed in chapter 12. The field instrument in the routine check phantom is compared with the secondary standard at the same depth in the water phantom in the x-ray beam and this yields a factor which takes into account the properties of the field instrument and the fact that the check phantom is not precisely water equivalent.

To use this system for routine checking of the dosimeter calibration it is necessary to establish that the calibration of the field instrument remains constant. This can be done by regularly observing its reading when exposed in fixed geometry with respect to a radioactive source (usually strontium-90) and correcting for the half-life of the source. The field instrument should also be regularly intercompared with the secondary standard at the energies for which it will be used and ideally on the machines for which it will be used.

The same check phantom can be used as a check on the flatness of the radiation field. The phantom is first centred in the widest radiation beam available and after a reference output measurement is made it can be moved, for further measurements, by fixed amounts in any direction normal to the beam by using the scales on the treatment couch. Quick flatness measurements can be made using this technique at any of the cardinal gantry angles.

A satisfactory calibration check, plus an acceptable ratio for the readings obtained in flatness measurements, are usually an adequate check that x-ray beam quality is constant. A change in beam quality introduces a mismatch between the beam profile and the flattening filter, which results in an unflat beam. If necessary a further check of beam quality can be made by turning the phantom over and measuring the output at a second depth. The ratio of this measurement and the one at the reference depth is a reasonable index of beam quality.

There is nothing very critical about the material chosen for the routine check phantom just described. Its purpose is to ensure that the radiation output remains consistent with the output determined more accurately at the time of commissioning. Its main requirement is that it is adequate, cheap and convenient. Any phantom having the necessary properties will have large thermal capacity and poor thermal conductivity and will therefore be slow in coming into thermal equilibrium with its environment. This is significant in the context of routine dosimetry checking because the ionization chamber placed in the phantom has to be at a constant temperature over the period of the measurements. It is convenient to have a set of phantoms, one for each machine, which are kept in the treatment rooms and are then always at the right temperature. This is quite practicable with a cheap material such as Perspex. Phantoms of the dimensions required are also quite heavy, typically 18 kg, and this is another reason for keeping them where they are to be used.

There are many advocates of multidetector systems, which have advantages for some of the functions outlined here, e.g. for checking field flatness. However these systems have the disadvantage that an indication of a non-uniform field, to the tolerance required, could be real or due to a fault on one of the detectors. The single-detector system is preferred because it avoids this difficulty, the only requirement being that its sensitivity should remain constant for the duration of the test. More sophisticated test equipment is very useful if extensive testing is required and particularly so if adjustments to operating conditions are necessary. A good example is the remotely controlled beam flatness scanner shown in figure 14.2, which can be attached to the accessory ring to facilitate flatness measurements at any gantry angle.

*14.2.2. Electron field measurements*

For electron field measurements the basic calibration is again carried out in water as discussed in chapter 12, while a polystyrene phantom similar to that in figure 14.1 can be used for routine dosimeter checks. In this case the depth of the ionization chamber hole will be nearer the surface to be

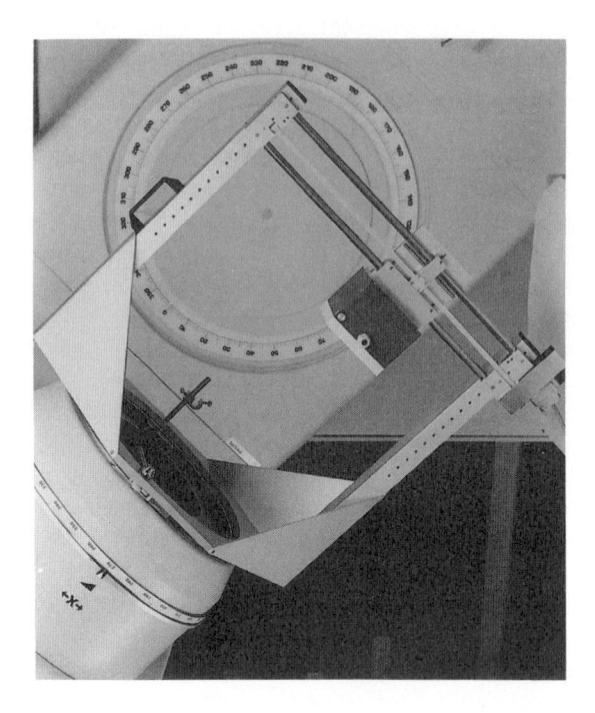

**Figure 14.2.** *An 'in air' scanner for beam flatness measurements.*

near to $d_{max}$ for each electron beam, e.g. centred at about 1 cm deep for 5 MeV electrons or 2 cm deep for 15 MeV electrons. It may be noted that, for either x-ray or electron beam measurements, the use of the ratio of the readings at two depths in a check phantom as a measure of radiation quality is essentially a constancy check. To use depth dose measurements in order to determine radiation quality definitively it is necessary to make measurements in a water phantom.

Galbraith *et al* (1984) have pointed to errors that can arise in electron beam dosimetry due to the accumulation of charge in a plastic phantom. In a conducting material such as water or a solid tissue substitute which has some electrical conductivity, this problem does not arise.

### 14.2.3. Detailed routine measurements

*14.2.3.1. Monitor calibrations.* This is the most important of the routine radiation checks and should be carried out several times a week, ideally daily. The routine check phantom is set up at 100 cm SSD in a 10 cm × 10 cm radiation field with the field instrument ionization chamber in its cavity. The readings for a group of exposures corresponding to one hundred dose

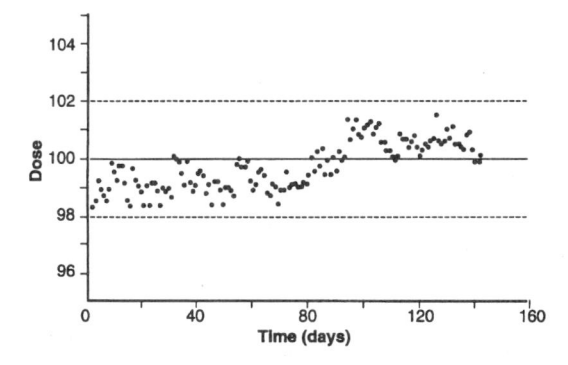

**Figure 14.3.** *A plot of calibration results for a linear accelerator over 144 consecutive working days. The results are expressed as dose per monitor unit under reference field conditions. $d_{max}$, 10 cm × 10 cm field at 100 cm SSD.*

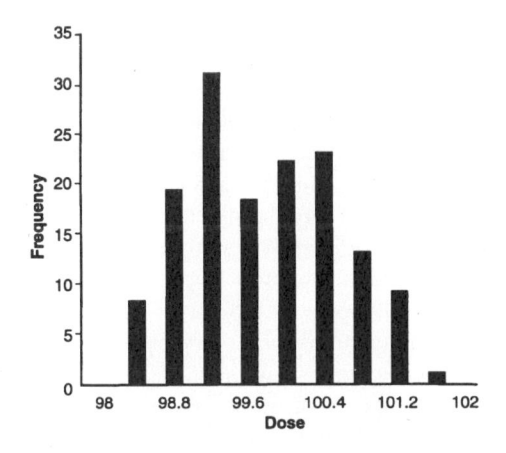

**Figure 14.4.** *A histogram of the data shown in figure 14.3.*

monitor units are noted, the mean value corrected to 760 mm Hg and 20 °C, and multiplied by the factors which give the measured dose in centigrays. Figure 14.3 shows the results of a series of such measurements for 4 MV x-rays over a period of 144 d, where each point is the mean of three readings. The range of the three readings did not exceed 1% in any group and any increase outside this value should be regarded as a possible indication of a fault on either the monitor or the field instrument. The variations shown in figure 14.3 are just about within acceptable limits. Figure 14.4 shows the frequency distribution of the calibration readings observed over the same period.

Although it is to be expected that any group of readings will show variations, those shown in figures 14.3 and 14.4 require some comment. There are four main sources of variation in the observed values in centigrays per hundred monitor units:

(i) variations in the dose distribution inside the radiation field and in radiation quality;
(ii) variations in the sensitivity of the monitor dosimeter system;
(iii) variations in the sensitivity of the field instrument;
(iv) variations in the way the system is set up for calibrations.

Of these four sources (iii) was written out at an early stage of the discussion, as regular testing with a radioactive source showed that the field instrument was constant within 0.5% over the period concerned in figure 14.4. (iv) is also not likely to be a serious contributor to the variations as geometrical errors in setting up the system for calibration (e.g. an error of 2 mm in the SSD) are unlikely to produce changes in the reading of the field instrument of more than 0.4%). There are also random errors on the values shown in figure 14.3. Since the individual points are each the mean of three readings with a range as stated, the random errors on these values as expressed by their standard deviation are less than 0.3%.

Although relatively elaborate systems are used to stabilize the radiation field, these are not perfect and so variations are possible inside the 1–2% range under discussion. Variations in the sensitivity of the dose monitoring system can arise from drifts in the electronics or because the monitor ionization chamber is not completely gas tight†. The atmospheric pressure can be subject to quite violent changes, especially under storm conditions, and it is difficult to eliminate or demonstrate very small leaks where the pressure in the ionization chamber may vary by a few per cent over several days in response to such changes. A gross gas leak will show up quickly as a fast response to changes in atmospheric pressure and temperature.

If the ionization chamber contains plastic material as electrical insulators, then over very long periods radiochemical reactions may result in the emission of gas and changes in pressure in an otherwise gas tight chamber. Good design will place the plastic insulators outside the nominal edge of the radiation field but, even under these conditions, they will still receive a sufficient radiation dose to produce observable radiochemical effects over long periods. This will cause a long-term drift in the monitor calibrations, which will normally be corrected by adjusting the sensitivity of the system.

In considering the results of dosimeter calibrations it is useful to lay down 'action limits' to deal with likely variations. In thinking about these limits it is useful to bear in mind that those patients for whom the radiation dose is

---

† This problem does not arise in an ionization chamber open to the atmosphere. In this case atmospheric pressure and temperature have to be measured and the sensitivity of the system automatically adjusted. The integrity of the pressure and temperature transducers is a prerequisite of a stable beam monitoring system.

most critical are being treated on a fractionated schedule and it is therefore sufficient that the calibration should average within 2% of its nominal value over the period of the whole treatment.

Suggested limits are the following.

(i) *Calibration within 2% of nominal value*: no action.
(ii) *Calibration within 2–5% of nominal value*: check the calibration fairly quickly, but otherwise take no action.
(iii) *Calibration more than 5% from nominal value*: repeat immediately with a different measuring instrument and reset the sensitivity of the monitor if necessary.
(vi) Look at the average value of the calibrations over a week or two, even if they have all met condition (i), and decide whether the monitor sensitivity needs to be reset to correct for small changes and slow drifts.

Any attempt to hold the calibrations within narrower limits are likely to involve many time consuming adjustments of the monitor dosimeter and may, in fact, make the average values over a period worse in relation to the nominal value.

Ideally routine monitor calibrations should not be carried at the same time each day. It is conceivable that the system could be going through a daily cycle of variations which would not be shown if the calibration was always done at the same time.

*14.2.3.2. Field flatness checks.* As already explained, the field instrument in the check phantom may be moved around the radiation beam to perform routine checks on field flatness. It is a matter of experience with any particular linear accelerator to decide how frequently these should be carried out. Although many machines are fitted with a field flatness monitor (see chapters 5 and 7) which is part of the beam stabilizing system, it is still worthwhile to carry out the independent checks as described. Provided the field flatness monitor gives a satisfactory reading, it is probably adequate to make field flatness checks about once a month. The readings should be within ±3% of those measured at commissioning to meet specification. Note that the optimum flatness of the widest beam will not necessarily be perfectly uniform, hence the check should confirm stability of the profile rather than absolute uniformity.

These comments apply to both x-ray and electron beam measurements.

*14.2.3.3. Beam quality checks.* Readings at two depths, by inverting the phantom or by adding additional material to its entrance surface, to check the quality of the x-ray or electron beams need to be performed regularly, again, probably on a once a month basis, and the mean value of the ratio of the readings should remain within ±1%.

*14.2.3.4. Wedge factors.* The use of wedge filters changes the relationship between the dose monitor reading and the dose delivered at the reference point in the radiation field. The dose in monitor units may be related to the dose at the reference point by means of wedge factors which are the ratio of the doses measured in the phantom in figure 14.1 for a constant number of monitor units, with and without a wedge filter in position. These factors are a slowly varying function of field size and are different for each wedge filter.

The value of the wedge factors clearly depends on the correct positioning of the wedge filter in the field, i.e. on the correct operation of the mechanical system which locates the wedge filters.

This can be conveniently and adequately checked by routine measurement of the wedge factors. Routine wedge factor measurements are only required at a single field size for each wedge as a check on the constancy of the factors. Again, monthly measurements are probably sufficient and the mean values of the wedge factors should remain inside $\pm 1\%$. If a universal wedge is used a single measurement of the wedge factor for the undiluted wedge is sufficient for all wedge angles.

*14.2.3.5. Rotation therapy measurements.* If the system is used for moving beam therapy it is necessary to check that the number of monitor units delivered per degree of gantry rotation remains constant for all setting likely to be used. This can be a self-checking parameter when the machine is in use, if the operators are instructed to report deviations of the rotation limits as prescribed for each treatment of more than an agreed limit, say $\pm 2°$.

*14.2.3.6. Field size measurements.* In chapter 12, a method of using a slow x-ray film and a rule to measure x-ray field size and symmetry about the isocentre and its relationship to the main optical beam was outlined. This provides a quick method of checking field size to an accuracy of a few millimetres, and should be carried out monthly, for a range of field sizes. A single sheet of film can be used to record field sizes of 5 cm × 30 cm, 10 cm × 10 cm, 15 cm × 15 cm and 30 cm × 5 cm. This enables measurements over a sufficient range for both the $X$ and $Y$ collimators.

*14.2.3.7. Shadow tray measurements.* The shadow tray interposes a Perspex plate, or other material, to support field shaping blocks between the x-ray head and the patient and this will slightly attenuate the x-ray beam. In other words, the relationship between dose monitor reading and the dose as measured under reference field conditions will be changed by the presence of the shadow tray. The dose in monitor units may be related to the dose at the reference point in the radiation field by the use of a measured 'shadow tray factor'. This is the ratio of the doses measured in the phantom in figure

14.1 with and without the shadow tray in position for a given dose. As was pointed out in chapter 6 the Perspex plate in the shadow tray needs to be changed regularly and the shadow tray factor will have to be remeasured when this change is made.

*14.2.3.8. General comment on routine radiation measurements.* All statements about dose to individual patients are ultimately dependent on the measurements just discussed. For this reason the execution of the measurements and the keeping of adequate records is important. These records should be kept for at least the period during which patients are kept on detailed follow-up, perhaps five years.

For a linear accelerator providing a wide range of radiation fields, a relatively elaborate schedule of daily and weekly radiation calibrations and checks is needed to cover all the radiation qualities used. The weekly check routine can be varied so that the items listed above come up about once a month. Unless this schedule is formalized in detailed work instructions, which are part of a formal quality assurance system, the tests are not likely to be performed properly.

This section has outlined some of the more important tests that are required on a routine basis and has suggested simple ways that they can be carried out. More detailed descriptions are available elsewhere including 'Commissioning and quality assurance of linear accelerators' *IPSM Report* 54 (1988) and chapter 5 of *Radiotherapy Physics* (Williams and Thwaites 1993).

### 14.3. ROUTINE MAINTENANCE AND SERVICING

The fact that equipment of such complexity needs to be regularly serviced hardly requires detailed justification. In any particular institution the basic decision to be made under this heading is whether to have the equipment regularly serviced by the manufacturer (or his agent) under contract, or to employ staff to do this locally. There are no hard and fast rules that can be used in making this decision, it must depend, in broad terms, on the availability of local staff or alternatively on the availability of an adequate commercial service.

For a hospital based service team it is probably more convenient, both from a service point of view and in relation to the actual treatment of patients, to carry out the servicing in short and frequent sessions; say in one day session per machine per month. For a contracted servicing arrangement it is more likely that the work will be done on a four times a year basis, the machine being out of clinical use for one or two days for each service. Some hospital based expertise will always be required to deal with unplanned maintenance resulting from minor breakdowns which can often be rectified in a matter of a few minutes.

Whichever servicing arrangement is in place some daily testing and servicing of the equipment will be required, some already dealt with under the heading of radiation measurements, some mainly mechanical, e.g. accuracy of front and back pointers and accuracy of optical beam devices.

Service personnel need to be well trained in the operation and servicing of the equipment and made aware of the major hazards, electrical, mechanical and radiation, involved. In other words, only trained personnel working to clearly defined schedules should be allowed to service the equipment. It is advisable that no one should be allowed to work on the equipment unless another competent person is present.

### 14.3.1. Some servicing details

*14.3.1.1. Mechanical.* A very high proportion of the work involved in servicing the equipment is mechanical: cleaning and lubrication, ensuring that nuts remain tight and inspection for mechanical wear. The tests on the mechanical stability of the isocentre and of the beam defining and beam direction systems need to be carried out regularly. The beam direction systems need to be looked at daily, as they are subject to more or less continuous handling and likely to suffer minor damage. The major items e.g. centring of the collimators, need to be examined in detail several times a year.

*14.3.1.2. Water cooling system.* The valves, flow switches and pipework of the water cooling system may include rubber or plastic components with a fairly limited life mainly because of the effects of the water itself. Those elements located near the treatment head may also be subject to long-term radiation damage. Most flow detectors operate by detecting the pressure difference across a metal orifice, and corrosion of the edge of the orifice will change their sensitivity. All these components need to be inspected several times a year and a schedule for regular component replacements is required for any particular machine.

The electrical elements in the flow valves, which protect the machine components from inadequate cooling need to be tested regularly as they will rarely be required to operate during normal operation. Even in the closed circuit water cooling systems described in section 4.3 there will be a tendency for solid and fibrous material to accumulate in the system due to corrosion of metal and non-metal components. This material will be prevented from circulating by the filter on the main pump line. This filter needs to be inspected, and if necessary changed, several times a year.

Where the local water supply is reasonably clear of solid matter and dissolved salts, it is useful to change the water in the main tank several times a year. Where these conditions do not apply it is important to have local discussions with the machine manufacturer about water quality control.

The refrigerator unit in the water control system, if one is used, is very specialized technologically in relation to the other systems in linear accelerators and is best serviced by an appropriate firm or agency.

*14.3.1.3. Vacuum systems.* In a continuously pumped machine the best policy is probably to keep careful records and leave the system alone in the absence of adverse signs. The most likely reason for having to re-pump the system is to change an electron gun, and this process is discussed in some detail in section 4.1.2. The second most likely reason for a vacuum failure is a leak in the thin window through which the accelerated electrons emerge from the vacuum system.

Ion pumps have a finite life because the gas pumped is accumulated in the sputtered material, which will eventually become saturated. The useful life depends on the pressure at which the system is operating. In a linear accelerator operating at a pressure of $10^{-7}$ Torr, the ion pump or pumps may need to be replaced about every five years.

*14.3.1.4. Electronic systems.* These divide into the high-power microwave system and its supply, the modulator and the control and dosimetry systems.

*The modulator and microwave system:* Again it should be mentioned that these parts of these systems operate at lethal voltages, some of which may be held for an appreciable time on capacitors after switch off. Although there should be built in safeguards in the system (e.g. interlock switches which cause the three-phase mains contactor to open when the modulator cabinet is opened and prevent further switching on until the cabinet is closed), it is essential to have well established procedures for those trained staff who are authorized to service the unit. It is in this context that it is important to ensure that two trained persons are present so that they can prompt each other about the procedures. The use of an earthing device on all high-voltage components (after switch off) is essential. The earthing device, or stick, is joined to earth by a heavy braid and should be a permanent part of the equipment, not just an *ad hoc* device used when someone happens to remember. The consequence of a 'double failure', that is failure by service personnel to switch off the modulator, plus a failure of a modulator interlock switch to open, is so lethal that it should be catered for by the safety system. If the interlock on a cubicle door is operated by a switch, then two door operated switches in series will still provide a safe system if one of the pair fails to operate.

There may still be mains voltage inside the cubicles after the safety interlocks have operated correctly. To protect service personnel, components carrying mains voltage, e.g. contactors, should be fitted with suitable shrouds or covers.

A special feature of this part of the equipment is that some of the high-powered valves, e.g. thyratrons and magnetrons, can only be tested in the system itself. If it is the policy to hold these spares locally they should be regularly changed for test purposes, rather than being held for long periods in an untested state.

*The control and dosimetry systems:* These low-powered electronic systems are best handled by a policy of keeping careful records and, otherwise, left alone in the absence of adverse signs. The routine calibration procedures discussed earlier in this chapter are basically a check on these systems.

In current electronic systems, where the circuit elements are usually mounted on printed circuit boards (PCBs), which may be connected to other circuits by edge or indirect connectors, electronic faults are usually best handled by changing a whole PCB and this implies that spares are carried in this form. These systems are rather sensitive to the presence of dust and can be cleaned by the use of a soft brush. Whether or not there should be a routine cleaning procedure depends on the local environment. While it is clearly useful to keep the PCBs clean, if this also involves disturbing the connectors the well intentioned maintenance activity may lead to other problems.

As is normal with any type of electronic equipment, the most vulnerable components are often electromechanical devices but, again, these are usually best left alone in the absence of trouble.

*14.3.1.5. Computer systems.* There is little by way of electronic engineering maintenance that is possible or desirable on computer systems. However, they require significant regular attention to ensure that data have not been corrupted and that they can be restored in the event of such corruption. It is good practice to maintain frequent back-up copies of all data, including programs, and to store these copies in a secure location away from the treatment machine. A complete back-up of the data from a complex computer controlled accelerator to magnetic media can take typically 30 min.

*14.3.1.6. Interlocks and limit switches.* There should be a routine for regularly testing interlocks and limit switches.

## 14.3.2. Record keeping

Since the useful life of a linear accelerator is likely to be up to 15 years, it is necessary to keep detailed records of machine performance and servicing procedures so that long-term lessons can be learned about the equipment and its operation.

The records of dosimeter calibrations should be kept separately as mentioned earlier, since they relate directly to patient treatments and may be called for, for example, if a patient shows an unexpected radiation reaction.

It is useful to keep the remaining performance records in the form of a daily log, where the values of measured machine parameters (for example, field flatness monitor reading, vacuum pressure) are entered on a suitable form. A record that such readings are at their normal values is useful in itself, while any small variations or drifts may be a sign of impending problems. For example, a drift in the electron gun filament current, required to maintain a flat beam, may be a sign of a coming failure.

The daily log should also record all faults, servicing activities, pre-planned or otherwise, and fitting of replacement components. It is useful to have integrating clocks recording the total time the machine is switched on and ready to produce radiation, and the total time when the machine is producing radiation. If the clock readings are entered on the log, then the useful lives of replacement components can be determined.

### 14.3.3. Electrical wiring

The complex systems of electrical wiring that have to pass from a fixed position into the rotating gantry and from the gantry into the rotating frame of the treatment head will accumulate minor mechanical damage over a long period, resulting in circuit breaks. Experience with machines built before 1970 suggested that a complete system may need to be replaced once in a machine used for 15–20 years. However more modern machines have proved much more reliable in this respect.

### 14.3.4. Spares

Policies about holding spare parts depend broadly on whether the equipment is serviced using local staff, or on contract by the manufacturer or his agent. In the former case it will be necessary to hold a fairly large stock of spares, though it will still be necessary to require the manufacturer to hold major items. For example, it would not normally be considered necessary to hold local spares of items such as major transformers, but it would be desirable to ensure that these are available from the manufacturer. As a rough guide, the initial stock of spares for a new machine, if held locally, would cost up to 5% of the capital cost of the machine. Annual consumption of spares is likely to be fairly modest, the main regular replacement items being the high-power valves in the microwave supply, i.e. the thyratron in the pulse forming system and the microwave valve.

If there is a maintenance contract which requires the manufacturer to hold and supply all spares, the location of these should be specified. It is worth remembering that, while air transport can move items half-way round the world in 24 h, this is only an advantage if the component can be dispatched from the supplier and delivered to the hospital in the same time scale. There are constraints other than the journey time in the supply chain!

228    *Accelerator operation*

*14.3.5. Long-term operation of linear accelerators*

As an example of the performance achieved by linear accelerators, it may be of interest to summarize the experience with the first four machines of this type used at the Christie Hospital, Manchester, and for the eight machines that were in use during the three-year period from 1993 to 1995. The performance data are summarized in tables 14.1 and 14.2.

The early machines summarized in table 14.1 were serviced by hospital staff, for one half day (a long half day running into the evening) every four weeks. In addition, running up and preparing the machines each day, and regular calibration and testing required about 1–2 man hours of activity per machine per day. About half of this daily work took place before scheduled start of patient treatments. These times are not included in the totals in table 14.1.

It will be noticed that the scheduled days out of service in the table are in whole numbers of five-day weeks. This arose because each machine was taken out of service at pre-planned times for major overhauls. For example, the machine installed in 1953 was taken out of use for two weeks when it was ten years old. Both the 1953 and 1962 machines used oil diffusion pumps, without a cold trap, in their vacuum systems, and tended to accumulate carbon on the inside surface of the accelerating waveguide, due to radiation effects on the low-pressure oil vapour. This reduces the efficiency of the guide, and during the two-week overhaul the carbon was removed. This problem did not arise with the 1968 and 1974 machines which used ion pumps. However, as a further example, the 1968 machine was taken out of use for two weeks at ten years old to replace all the cables which rolled and unrolled as the gantry rotated.

Unscheduled days out of service in table 14.1 are defined as days when it was not possible to treat patients because of a machine failure. In addition to these there were of course numerous occasions when machine faults were quickly rectified by hospital staff so that patient treatments could continue.

It should be pointed out that the early linear accelerators mentioned in table 14.1 were very simple machines by current standards and that this accounts for their very few days out of service. The 1962 machine had a working life spanning the period of transition from valve electronics to solid state electronics, and its control and dosimetry circuitry was modified to take account of this. Multipurpose machines, such as the 1968 model, are necessarily a good deal more complex and this offers more possible causes for breakdown, and greater difficulty in diagnosing faults.

The current generation of linear accelerators is very much more complex than those in table 14.1, mainly because they are more versatile and incorporate much more elaborate safety, automatic control and computer controlled operating systems. They include many more components and circuits which may develop faults and it is perhaps surprising that these later machines have a similar record of down time as shown in table 14.2.

**Table 14.1.** The first four linear accelerators used at the Christie Hospital, Manchester.

| Machine | Installed | Scrapped | Scheduled days out of service | Unscheduled days out of service | Years in use | Average unscheduled days out of service per year |
|---|---|---|---|---|---|---|
| 4 MV x-ray generator | 1953 | 1971 | 10 | 19 | 17 | 1.1 |
| 4 MV x-ray generator | 1962 | 1984 | 20 | 23 | 22 | 1.05 |
| 8 MV x-ray/3–10 MeV electron generator | 1968 | 1984 | 20 | 30 | 16 | 1.9 |
| 4 MV x-ray generator | 1974 | 1987 | 5 | 28 | 13 | 2.1 |

**Table 14.2.** *Eight linear accelerators in use at the Christie Hospital, Manchester, 1993–1995.*

| Machine and accessories | | Installed | Years in use | Average unscheduled days out of service per year | Average downtime (faults exceeding 15 mins) (%) |
|---|---|---|---|---|---|
| 6 MV x-ray generator | | 1993 | 2 | 0.5 | 0.58 |
| 6/8 MV x-ray/4–15 MeV electron generator | MLC (1995) | 1993 | 2 | 1.5 | 2.2 |
| 4 MV x-ray generator | | 1994 | 1 | 2.5 | 1.7 |
| 8 MV x-ray/4–14 MeV electron generator | | 1982 | 13 | 0.5 | 1.1 |
| 4 MV x-ray generator | | 1983 | 12 | 2.0 | 1.6 |
| 8 MV x-ray/4–14 MeV electron generator | | 1986 | 9 | 0.75 | 0.5 |
| 4 MV x-ray generator | EPID (1993) | 1986 | 9 | 2.75 | 1.7 |
| 6/20 MV x-ray/4–22 MeV electron generator | MLC (1989) EPID (1990) | 1987 | 8 | 1.75 | 2.8 |

As with the early machines they have been maintained by hospital staff apart from some activities carried out in the warranty periods. The data for these machines are presented in a slightly different form. The unscheduled days out of service include only periods of one half day or more, as it is in these circumstances that there is major disruption to the department's schedule. However all faults requiring more than 15 min to correct have been recorded and the average down time attributed to such faults is expressed as a percentage of the total time for which the machines were scheduled to be available. It has not been the policy to take these machines out of service for major overhauls as was the case with earlier machines but they have been subject to scheduled periods out of service for upgrades, such as the installation of the MLCs and EPIDs.

Any complex and successful equipment goes through a fairly well defined set of development stages during its working life, and linear accelerators are no exception in this respect. If a machine is a new model it will go through a phase when minor design faults are gradually corrected. Even if it is of a well established design, a new machine is likely to have excess breakdown time in its first few years of use due to minor manufacturing defects or the use of components which are near the edge of tolerance. Also, during this stage, the staff who operate and service it will be becoming more proficient in its use and at diagnosing and correcting faults. At the end of this period the system should operate at optimum reliability, when components with a limited life are regularly replaced. It appears from these limited data that the 8 MV machines installed in 1982 and 1986 are more reliable than the simpler 4 MV x-ray machines. This is attributed to the need to replace electron gun filaments on the lower-energy machines (which run at higher beam currents), a procedure which usually takes more than half of one day.

This level of performance can be achieved only by having an adequate complement of well trained engineering staff, by investment in reasonable stocks of spare parts and by co-operation of the manufacturers in providing advice and additional spare parts at very short notice. In addition the co-operation of radiotherapists and radiographers in ensuring that sufficient time is set aside for planned preventive maintenance is essential.

### 14.3.6. *General comment*

Safe operation of the system, either during the test and servicing procedures discussed in this chapter or when the machine is in clinical use, depends ultimately on the people who use it. This requires adequate training of all personnel and on their subsequent supervision. Perhaps the most important safety element in the whole operation is to ensure that at any particular time a specified person is in charge of the equipment and that no one enters the treatment room or operates any part of the machine without the permission of this person. When the machine is in clinical use this person will normally be a nominated senior radiographer. When the machine is in use for calibration, testing or servicing purposes, either a physicist or an engineer should be nominated as the person in charge.

Linear accelerators have proved to be very successful as radiotherapy treatment machines for three main reasons.

(i) They give well defined and stable radiation beams with acceptable penetration properties.
(ii) With proper servicing, unscheduled time out of service due to machine failures can be limited to two or three days per year.
(iii) High utilization can be achieved, resulting in a surprisingly modest treatment cost per patient as will be shown in the following section.

The useful working life of a well serviced linear accelerator is likely to be 10–15 years and is usually determined by technical developments and obsolescence of components. Obsolescence has been a particular problem for computerized systems which are extremely reliable but which rely on peripheral devices and consumables whose commercial product lifetime are often much less than would be expected for a linear accelerator.

### 14.4. OPERATING COSTS

This section outlines the calculation of the average annual cost of operating a linear accelerator at 1996 UK prices.

The main elements are given in table 14.3 which is in the form of a work sheet giving the total operating cost over 10 years for two linear accelerators, a simulator and a treatment planning system. The first accelerator is a 6 MV x-ray machine. The second is a more versatile machine providing x-rays at both 6 and 12 MV and electrons in the range from 4 to 15 MeV, and fitted with an MLC and an EPID. Interest on the initial cost of the machines and their optional accessories is calculated on the assumption that their value is fully depreciated in equal increments over the 10 years and that interest is charged on the diminishing capital at a rate of 6% per year.

Only one-quarter of the building cost is included as buildings are depreciated over 40 years. As with the equipment the annual cost is calculated taking into account interest charges of 6%.

It is assumed that a modest initial stock of spare parts cost between 2 and 3% of the capital cost of the treatment machines and simulator.

The main consumable items for the accelerators are magnetrons at a replacement cost of £5000 per year. The simulator is likely to need a replacement x-ray tube and a replacement image intensifier once during its 10 year life. Replacement parts for the treatment planning system are assumed to be covered by the maintenance contract. The cost of electricity is surprisingly modest and is calculated on the basis of the treatment machines

**Table 14.3.** *The cost of operating various items of radiotherapy equipment for 10 years.*

|  | 6 MV[a] | 12 MV[b] | Sim[c] | TPS[d] |
|---|---|---|---|---|
| Capital costs |  |  |  |  |
| Machine (fully depreciated) | £375K | £510K | £280K | £120K |
| Multileaf collimator |  | £200K |  |  |
| Portal imaging |  | £85K |  |  |
| Interest payments on equipment | £113K | £239K | £0.84K | £36K |
| Building (10 year depreciation —40 year life) | £90K | £90K | £25K | £12K |
| Interest payments on building | £76K | £76K | £21K | £10K |
| Initial stock of parts | £10K | £20K | £5K |  |
| Replacement parts/ consumables | £40K | £80K | £70K | £5K |
| Energy to operate machine | £5K | £5K | £5K | £5K |
| Building maintenance costs | £47K | £47K | £23K | £23K |
| Maintenance contract |  |  |  | £96K |
| Total equipment costs | £756K | £1342K | £508K | £302K |
| Cost per year | £75.6K | £134.2K | £51K | £30.2K |
| Cost per fraction | £7.6 | £13.4 |  |  |
| Cost per patient | $N \times$ £7.6 | $N \times$ £13.4 | £20.4 | £12.1 |

[a] 6 MV x-ray generator.
[b] 6 MV and 12 MV x-ray generator with 4–15 MeV electrons.
[c] Simulator.
[d] Treatment planning computer system.

**Table 14.4.** *Total costs (£) for delivering a twenty-fraction course of treatment including simulation and planning. (Add £250 for mould room preparation if required.)*

|  | 6 MV x-rays | 12 MV x-rays and electrons |
|---|---|---|
| Machine costs |  |  |
| Accelerator | 152 | 268 |
| Simulator | 20 | 20 |
| Planning system | 12 | 12 |
| Staff costs |  |  |
| Physics | 47 | 47 |
| Radiography | 200 | 200 |
| Total | 431 | 547 |

being switched on in their stand-by state for 10 h d$^{-1}$ and operated at full power for 2.5 h d$^{-1}$. The cost of heating and lighting, which is included in the building maintenance costs, is approximately double the energy costs for the machines themselves.

It is assumed that maintenance on the accelerators and simulator is carried out in house and this cost is identified in table 14.4. If a commercial maintenance contract is used the annual cost will be typically 8% of the capital cost of the equipment: this is the figure used for the treatment planning system. As computer control and networking becomes more common the costs of software maintenance are likely to become unavoidable and will need to be taken into account when planning for, and comparing the costs of, replacement equipment.

In relating the total cost of operating a linear accelerator for one year to the cost of treating a patient, the prime index is the cost per fraction. This is calculated on the reasonable assumption that a machine can deliver 10 000 individual fractions per year. The cost per patient can then be determined from the number of fractions required, $N$, which depends on clinical practice.

It is assumed that a simulator and a treatment planning computer can service a department treating 2500 patients per year (although not all patients will be simulated or planned) and this figure is used to calculate the average equipment costs per patient for these activities.

Table 14.4 gives the total costs for delivering a course of treatment consisting of twenty fractions. It includes staff costs for physics, engineering and radiography as well as the equipment costs brought forward from table 14.3. The cost of radiography includes treatment delivery, simulation and management and supervisory tasks within the department. Both these staff costs include overheads above the actual staff salaries. In so far as patient treatments may vary considerably in complexity, there are corresponding variations in treatment cost per patient. The figures in table 14.4 are average values based on known annual costs and numbers of fractions delivered in a year. Some patients will require mould room preparation, which might include production of a treatment shell, or customized shielding blocks. An average of about £250 should be added to the cost for the treatment of such a patient.

These costs should be treated with some caution. They are based on experience in a large radiotherapy centre treating more than 10 000 patients per year on eight linear accelerators. At this level there are many economies of scale and it unlikely that the equivalent costs in smaller centres will be less than these.

It is important to note that these costs are for treatment delivery. They do not include medical staff costs nor the costs of hospitalization and other diagnostic procedures such as CT or MR scanning. No account is taken of inflation or any variations in interest rates over the 10 year period.

# CHAPTER 15

# SIMULATORS AND TOMOGRAPHIC SCANNERS

In this book about linear accelerators the emphasis has been on the delivery of treatment. However before treatment can be given it is necessary to carry out a range of tasks which come under the heading of treatment planning, in its most general sense. These include diagnostic tests to identify and localize the site to be treated and just as importantly the structures to be shielded, simulation of proposed treatment to ensure that it will achieve the required results and finally modelling of the treatment, usually in a computerized treatment planning system in order to design a dose distribution which the radiotherapist knows from his experience will provide the best compromise between tumour control and morbidity.

## 15.1. SIMULATORS

A simulator consists of a diagnostic x-ray source mounted in such a way that it can provide a radiation field to simulate the field used in a radiotherapy treatment. If the device is to simulate the treatments given on a linear accelerator, and this is usually the case, then the mounting for the x-ray source and the patient support system will be designed to form an isocentric system, according to the principles and specifications given in chapters 6, 8, 9 and 12. A simulator is used to set up a patient as for radiotherapy, to make an initial determination of field size required and to use the diagnostic quality x-ray field to verify that the desired anatomical volume will be irradiated. To visualize the x-ray field, the simulator normally includes an image intensifier as well as facilities for taking radiographs. The purposes and developments of radiotherapy simulators are discussed in the *British Journal of Radiology Supplement 23* (1989).

The basic features of a simulator are shown in figure 15.1(a) and a photograph in figure 15.1(b). The x-ray unit and image intensifier are mounted on a U-shaped structure so that they remain in line when it is

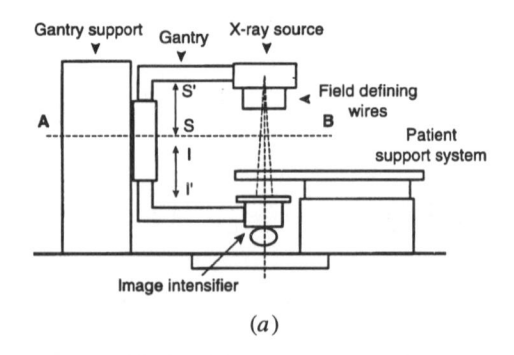

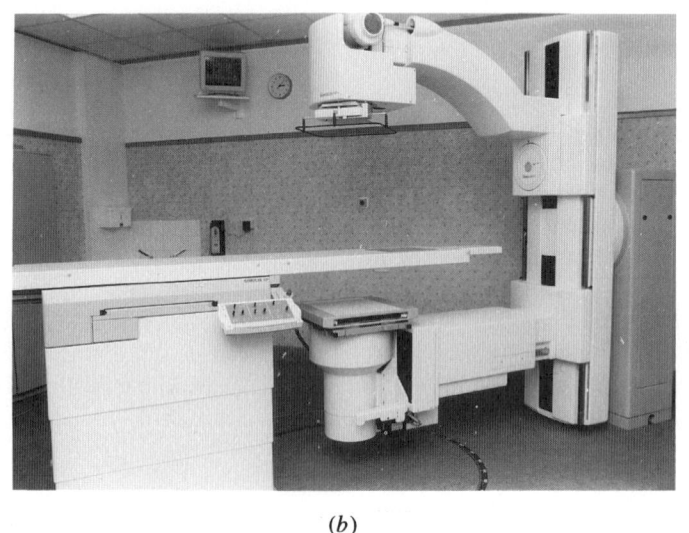

**Figure 15.1.** (a) A diagram to show the main components of a simulator. (b) A photograph of a simulator.

rotated about the horizontal axis AB. The field size is simulated by two orthogonal pairs of wires representing the four independent collimators in a linear accelerator's treatment head. The projected image of the wires allows the anatomy inside and outside the proposed radiotherapy field to be shown. The field defining wires are mounted such that they rotate about the central axis of the x-ray field and, as with the linear accelerator, this axis, the axis of the gantry rotation, and the vertical axis of rotation of the patient support system all pass through the isocentre. Image intensifiers do not have a large enough field of view to cover the largest radiotherapy fields to be simulated (e.g. whole trunk). It is therefore necessary to be able to move the intensifier in figure 15.1(a) in two dimensions, parallel and perpendicular to the line AB so that these large fields can be explored. If the image from the intensifier is digitized then it is possible to build up and display a composite image

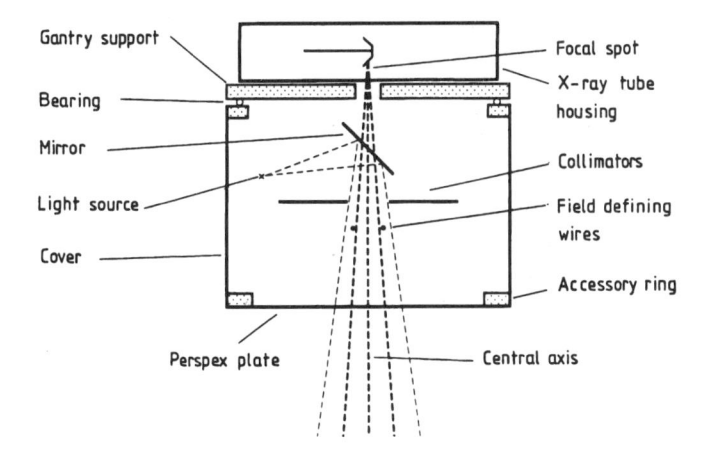

**Figure 15.2.** *The simulator x-ray head.*

covering the largest fields available from the accelerator. In the absence of such a facility large fields can be imaged only by film based radiography rather than fluoroscopy.

The simulator needs to have the same range of beam direction devices as the radiotherapy machine, i.e. optical beam, optical pointers, SSD indicator and mechanical front and back pointers, and these again need to have the same properties and specifications as those used on the radiotherapy machine. Since it is necessary to simulate irregularly shaped fields, a shadow tray is also required. It is likely that methods of simulating MLCs will be added to simulators as MLCs come into general use. These will probably be computer generated graphical representations overlaid on the digitized image intensifier image.

### 15.1.1. The x-ray head

Figure 15.2 shows some further details of the x-ray field defining system. The diagnostic x-ray tube is mounted on the gantry arm in its conventional tube housing so that the focal spot is on the line through the axis of symmetry of the field defining wires and the isocentre. The position of the x-ray tube housing needs to be adjustable in two dimensions to line up the focal spot to this requirement. The adjustable x-ray collimators then give the maximum field size required for a particular patient. The movable pairs of wires, about 1 mm in diameter, which define the required field for the therapy treatment, the x-ray collimators and the optical beam system all rotate about the central axis of the x-ray field on the bearing shown. A suitable mark on the Perspex plate shown will then give an indication of this axis in the optical beam.

The x-ray head also needs to carry a mounting for beam direction and beam shaping devices: this is labelled 'accessory ring' in figure 15.2.

*15.1.1.1. Use of the shadow tray.* The shadow tray should be suspended from the x-ray head at the same distance from the radiation source as on the therapy machine. In practice, if a simulator is used to simulate treatments to be delivered on therapy machines of different designs, alternative shadow trays or a single adjustable shadow tray are needed. Shadow blocks to shield a particular organ, or to produce a complex shaped field (see chapter 9), can be carried on this tray. The optical beam will then show that part of the patient shielded by the block and, if this is satisfactory, the outline of the field defined by the block can be marked on the patient's skin. The x-ray source and image intensifier may also be used to look at the field in relation to the patient's anatomy. The shadow blocks used on the simulator do not need to be as thick and heavy as those used on the treatment machine.

*15.1.1.2. The x-ray tube and generator.* Standard diagnostic equipment is used, the specification being largely determined by the need to simulate lateral pelvic fields. This will require a tube and generator operating at voltages up to 125 kV and tube loadings up to 400 mA s. The x-ray tube must have a small focal spot to achieve good definition of the field defining wires. These are necessarily closer to the target than the patient and therefore any blurring due to the focal spot size will be magnified. Fine and broad focal spot sizes of 0.4 and 1 mm are required, both of which are considered small in the practice of diagnostic radiology. The maximum tube loading quoted can only be used on broad focus. Field coverage up to 40 cm × 40 cm at 1 m is needed to simulate the largest radiotherapy fields available and this requires an x-ray tube with a target angle greater than 20°, somewhat greater than for a standard radiographic tube. The critical image requirement is to be able to see the field defining wires in a lateral pelvic projection during fluoroscopy.

### *15.1.2. The image intensifier*

The only special requirement for the image intensifier arises from the use of very large fields, as already mentioned. Most simulators in current use are fitted with 23–38 cm diameter electrostatic image intensifiers. The image from the image intensifier is observed by a TV camera and displayed on a TV monitor or monitors in the usual way. As mentioned in section 15.1 there are advantages in digitizing the image from the image intensifier. These include electronic storage, field of view extension (by adding images as the intensifier is panned across a large field) and image transfer to other computerized systems such as planning systems and electronic portal image work stations.

For recording radiographic images, arrangements are usually provided for mounting standard grids and cassettes in front of the image intensifier.

## 15.1.3. The patient support system

The requirements of the patient support system are exactly the same as those on the treatment machine and have therefore already been dealt with in section 9.3. However the presence of the image intensifier (see figure 15.1) limits the range of vertical movement of the couch so there is no need to have a system giving maximum vertical movement, as illustrated in figure 9.4.

## 15.1.4. The gantry

The limitations on couch movement just referred to determine the maximum field size that can be obtained by increasing the SSD for a fixed source to isocentre distance. This limitation can be removed by making the source to isocentre distance variable as indicated by the arrows SS' in figure 15.1(a). This also allows the system to simulate machines operating at source to isocentre distances different from 1 m.

It is also necessary to be able to move the image intensifier radially with respect to the isocentre so that it can be brought as close as possible to the patient. This movement is indicated by arrows II' in figure 15.1(a).

## 15.1.5. Control systems

The systems for controlling the x-ray generator and the image intensifier are standard. Some simulator manufacturers produce the mechanical systems that make up a simulator and leave it to the user to choose what x-ray tube, x-ray generator and image intensifier will be used.

Because of the need to vary the source to isocentre and the image intensifier to isocentre distances, the movements of the gantry on a simulator are more complex than those on the linear accelerator. When the source to isocentre distance is set to its standard value of 1 m the gantry can rotate through 360°. When this distance is increased the gantry can strike the floor. This can be avoided by arranging for an interlock switch to be operated when the system leaves its 1 m source to isocentre position, which will bring suitable limit switches into the movement control circuit.

The movements and controls on the patient support system are in principle the same as those on a linear accelerator.

The movements in the x-ray head, position of the beam defining collimators, position of the field defining wires and rotational movement of the whole system need to be controlled and read out to the same precision as the corresponding movements on the treatment machine. When the x-ray source to isocentre distance is increased from its normal value, the x-ray head may be out of reach of an operator working at floor level, so it is necessary to have all the moving parts motor driven, and their controls and read-out displays in a convenient position. The controls and read-out displays for all the moving parts of the machine need to be adjacent to the patient on the couch and may be mounted on the couch itself or on a mobile control unit. It is also useful to have these controls and read-outs duplicated in the shielded control area from which the x-ray generator is operated. The possibility of a collision between a patient and the x-ray head, or between the gantry and the couch assembly, as discussed under the heading of patient safety in section 9.6, is just the same for a simulator as for the treatment machine and can be dealt with using the same principles as outlined in that chapter. There is the additional possibility of a collision between the image intensifier housing and the patient support system and this can be dealt with by the use of a 'touch ring' or proximity switch system.

### 15.1.6. Simulator operation

Patients receiving radical radiotherapy treatments are treated daily over a period of several weeks, while patients receiving palliative treatment may receive their radiation dose in a single session. On average a course of radiotherapy requires the patient to be set up on the machine a large number of times. A simulator is used to try out the proposed radiotherapy treatment on a patient, so it is likely that a patient will be examined on it only once or twice. It follows that one simulator can deal with the patients treated on a number of linear accelerators, perhaps as many as four.

When a patient is set up as for treatment by a particular field on the simulator, then not only can the treatment field be visualized, as already discussed, but all the machine parameters (except those involving the dose monitoring system) can be read out for transfer to the treatment machine. Transfer can be manual or, in the case of computer controlled linear accelerators, electronic. When electronic transfer is employed the simulator can be considered either as a remote terminal to the accelerator or as an additional machine on a network of radiotherapy equipment which could include accelerators, treatment planning systems and CT scanners as well as the simulator.

### 15.1.7. Design of the simulator room

The functions of the simulator require it to have dimensions and movements similar to that of a linear accelerator. The geometrical considerations that determine the minimum size for the simulator room are thus very similar to those discussed in chapter 13 in relation to the linear accelerator. The radiation protection requirements are those for the diagnostic x-ray source used. The shielded area from which the x-ray exposures are controlled, and from which the movements of the simulator may also be controlled and monitored may be an adjacent room, with a lead glass viewing window, or a suitably protected cubicle inside the main room.

## 15.2. COMPUTED TOMOGRAPHY

Although the radiotherapy simulator is an essential tool it is not ideal as the images are simple radiographic projections along the proposed beam directions. Computed tomography provides much more detailed anatomical information, which, in many cases, is necessary for the delineation of the target volume and which can be used as input data for dose calculation in a computerized treatment planning system. A detailed description of the production of a computed tomographic (CT) section is clearly beyond the scope of this book but can be found in *The Physics of Medical Imaging* (Webb 1988) and further details of the applications to radiotherapy can be found in *The Physics of Three-Dimensional Radiation Therapy* (Webb 1993).

### 15.2.1. X-ray CT scanners

The introduction of x-ray CT scanning in the 1970s was important in removing several constraints which limited the accuracy of radiotherapy. Until that time precise three-dimensional localization of tumours and critical structures was difficult and the position, size and density of internal structures, such as the lungs, could only be estimated. CT scanners have developed mainly in response to the needs of diagnostic radiology and it is arguable that machines of highest imaging specification are not necessary for use in radiotherapy planning. However there are some aspects of performance where the needs for radiotherapy application are different from those of diagnostic radiology.

The aperture, into which the patient is placed, in a scanner used for radiotherapy planning must be as large as possible so that the patients can be imaged in the position in which they will be treated. Clearly a change in position can alter the geometric relationships between internal structures and the external markers which will subsequently be used during set-up for treatment.

For the same reason, as most treatment couches have a flat surface, it is necessary to provide a flat couch top for the CT scanner.

It must be possible to acquire sections in tilted planes, i.e. not normal to the long axis of the patient, so that treatments in these planes can be modelled in the treatment planning system. (Alternatively tools for manipulation of data collected in a set of normal planes can be used to reformat the data into the required planes.)

The CT scanner should have known accuracy in terms of geometric positional and density measurement for each volume element in the three-dimensional data set. A geometric accuracy of 1 mm is necessary and, if the actual measured density of each volume element is used in the dose calculation process, a densitometric accuracy of 2% is required. In order to make use of the CT data it is necessary to provide facilities for transfer to the treatment planning computer. This can be via removable media such as magnetic tape or optical disk or electronically via a network connection.

### 15.2.2. Production of computed tomographs—simulator based CT scanning

The production of a computed tomograph of a patient involves taking a large number of projections in many directions through the patient in the plane of interest, and back-projecting them. If the resultant image is to show the whole cross section, the x-ray field has to be sufficiently wide to cover the full width of the patient. The length of the field in the other direction (i.e. in the direction of the long axis of the patient) is restricted to not more than a few centimetres to minimize the amount of scattered radiation in the projections and more importantly to avoid blurring by the inclusion of structures whose cross section varies within the slice thickness. A system to carry out these processes needs to be able to produce and store the projected images, to back-project them to form the image of the tomographic slice and to display it. This is a very brief and crude outline of the processes involved in producing a CT image.

These functions can be performed on a simulator to which additional equipment has been added to detect, store and process the transverse projections. Two methods of detection have been employed, those using the image intensifier which is part of the simulator and others which employ an additional detector specifically for the collection of profiles for CT.

The limitations in the field size of the intensifier make it impossible to subtend the full width of the patient in a single projection. However the tomographic field of view can be extended to cover the patient. This is achieved by moving the x-ray field and intensifier off centre and rotating the gantry through 360°, as shown in figure 15.3. To limit the dynamic range of the information arriving at the image intensifier, a filter whose shape is roughly indicated in figure 15.3 is placed between the x-ray source and the patient.

The transverse projection images collected when the gantry is rotated through 360° are initially in the form of the analogue voltages which normally generate the video image seen on the TV monitor. By selecting the information from a set of adjacent raster lines the 'slice width' of the image may be chosen, and, by separating out the information from different sets of raster lines, several slice images may be formed. The video voltages along the selected lines may be digitized and from there on the storage, processing and handling are the same as for a diagnostic CT scanner. However the image quality is poor when compared with a state of the art diagnostic scanner. There are many reasons for this, including the lack of post-patient collimation, designed to reduce the influence of scattered radiation, and the relatively poor signal to noise ratio on the profiles, measured with a detector which has not been designed for this purpose.

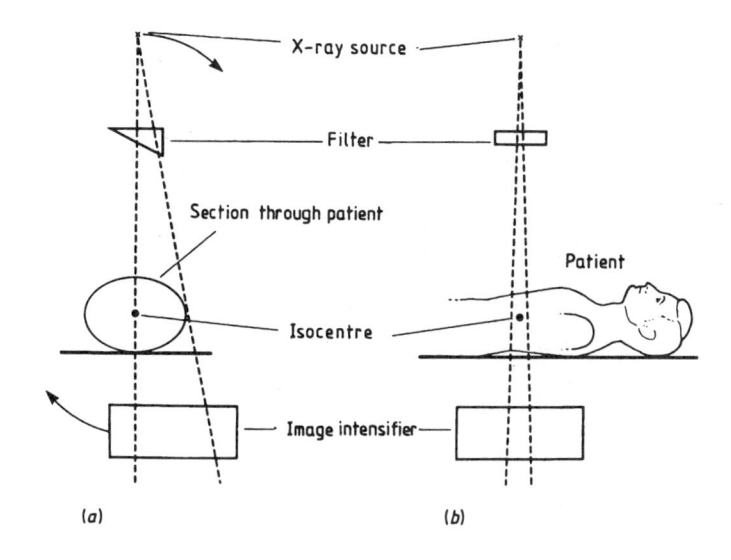

**Figure 15.3.** *The use of a simulator to produce a computed tomograph. (a) A section through the isocentre in the plane of rotation. The gantry is rotated through 360° to collect a full set of projections. (b) A patient in position to be scanned (lateral view, normal to (a)).*

The alternative method of detection is to add a linear array of solid state detectors to the front of the image intensifier, that is to use detectors which are similar to those employed in CT scanners. The advantages of this method arise from the use of collimation to reduce scatter and the use of detectors with higher efficiency and better signal to noise ratio.

Use of the simulator as a CT scanner places additional demands on the x-ray tube. This needs to be operated at high power and with a high duty cycle so that scanning of consecutive slices can be carried out without delays for cooling.

These methods of producing CT sections are still under active development and although, as has been stated above, the quality of the images is not as high as for a dedicated CT scanner they are sufficient for providing quantitative image data for the purposes of treatment planning. An advantage in using such a system rather than a standard CT scanner is that it provides the ability to scan patients on the simulator couch in the actual treatment position. This is because the effective aperture of a simulator based CT system is much larger than the 70–75 cm aperture on a standard CT scanner.

Figure 15.4 shows examples of images of sections through the head and the lungs generated by a simulator CT system using a solid state detector array.

(a)

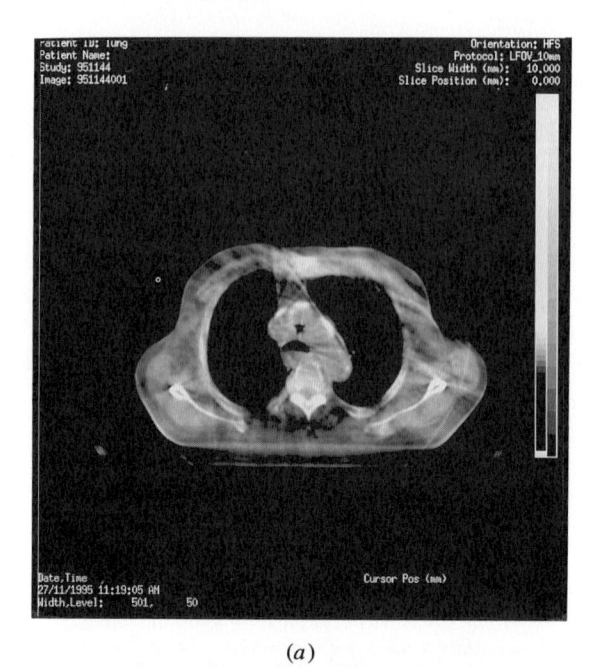

(b)

**Figure 15.4.** *CT images taken with a simulator. (a) A section through the head. (b) A section through the thorax. (Courtesy of Elekta Oncology Systems.)*

## 15.2.3. Magnetic resonance imaging—MR scanning

Although magnetic resonance imaging produces tomographic slices, with the possibility of greater discrimination between normal and abnormal tissues, the use of MR for treatment planning is not as straightforward as the use of x-ray CT. Firstly, the geometric accuracy of an image produced by an MR scanner depends on the uniformity of the magnetic field. This will not be perfect towards the periphery of the field of view so the possibility of geometric distortion arises particularly in the representation of the patient outline. Secondly, the pixel values (grey levels) depend on the magnetic characteristics of the tissues being imaged and the pulse sequences used in the imaging process. There is no unique relationship between the pixel value, tissue type, density or electron density. This means that the pixel values cannot be used in the treatment planning process for inhomogeneity corrections as is the case for x-ray CT. However, magnetic resonance is a very important imaging tool in the assessment and delineation of tumours. As a result methods of combining information from both CT and MR have been developed and the two types of imaging are seen as complementary rather than alternatives.

## 15.2.4. CT simulation—virtual simulation

If a series of contiguous CT sections is acquired covering a region of interest in a patient the three-dimensional data set produced can be manipulated to model the simulation process. Tumours can be delineated, virtual beams can be projected through the data set and digitally reconstructed radiographs can be produced to provide an image equivalent to the image that would be produced on a conventional simulator. These manipulations of the data can be carried out while the patient is still in position on the CT scanner so that, after decisions have been made about beam directions, field sizes, entrance points etc, reference marks can be applied to the patient for subsequent alignment on the treatment machine. Some CT machines that have been specifically modified to be CT simulators include optical devices which project the beam centres and outlines onto the skin surface for marking out. There are several potential advantages of virtual simulation. These include more accurate tumour localization and the ability to investigate the beam's intersections with structures which can be displayed in sagittal, coronal, transverse or any other plane. In addition the volumetric data set is precisely that needed for the dose calculation part of the treatment planning process. The disadvantages are that a dedicated CT scanner is required and that for acceptable resolution in all three dimensions it is necessary to take a very large number of thin contiguous slices, maybe 1–2 mm wide. Furthermore the provision of a CT simulator is unlikely to remove the need for a conventional simulator as the CT process is time consuming and therefore not acceptable for the simulation of simple treatments, which account for a large part of a radiotherapy department's work.

# CHAPTER 16

# CONTEMPORARY DEVELOPMENTS

Although linear accelerators have been used for radiotherapy for more than 40 years the technology and the applications have continued to develop. This final chapter outlines some of these applications.

## 16.1. DYNAMIC THERAPY

### 16.1.1. X-ray rotation therapy

The simplest form of dynamic therapy, which has been alluded to in earlier chapters, is rotation therapy, otherwise described as arc therapy or proportional arc therapy. While this form of therapy pre-dates the linear accelerator, and therefore can hardly be considered as a contemporary development, it is a logical introduction to newer forms of dynamic therapy. The principle underlying rotation therapy is that a high dose can be delivered at the intersection of multiple beams in the target volume while lower doses are delivered to tissues outside the target, where the beams do not intersect. In the case of two intersecting beams (or two orthogonal opposing pairs of beams) the dose outside the target volume will be approximately 50% of the dose in the target and for three fields it will be approximately 33%. Increasing the number of fields reduces the dose outside by spreading the energy deposited over a larger volume of the patient. The easiest way to achieve intersection of the beams in the target is to place the centre of the target volume at the isocentre and to make use of the isocentric rotation of the gantry to direct multiple beams which will then intersect at the target. Continuous rotation of the gantry during irradiation is equivalent to an infinite number of beams and minimizes the dose outside the target volume. In the simple examples above, the dose outside the target appears to be approximately inversely proportional to $N$, where $N$ is the number of beam directions. Unfortunately this not a general rule and three is probably the highest value of $N$ for which it applies. As $N$ increases most points outside the target volume will be irradiated by more than one beam (although not by all beams) and the doses will be higher than for the simple cases where tissues outside the target are only irradiated by one beam. For a very small beam the radial variation of dose outside the target volume is approximately $1/r$ (ignoring attenuation) and the dose near the surface is less than the target dose by a factor which is the ratio of the patient's circumference and the field size. Note that the $1/r$ function is the blurring function in CT scanning and is 'removed' from the blurred image formed by simple back-projection by one of the various image reconstruction algorithms. This will be referred to again in subsequent sections.

As far as the linear accelerator is concerned rotation therapy requires that the gantry should rotate in a highly controllable way during irradiation. In particular the dose monitoring and control system (chapter 7) have to be linked to the gantry control system (chapter 9). In proportional arc therapy the gantry rotation speed is proportional to the dose rate. This can be achieved by treating either the dose rate or the gantry speed as the independent variable. In the first case the dose rate is used to control the gantry speed; in the second the gantry is rotated at a constant speed and the dose rate controlled as the dependent variable. Clearly both the dose rate and the gantry speed can only be varied within limits and therefore it is sometimes necessary to restrict the independent variable to ensure that the dependent variable can be adequately controlled. For example, if a small number of monitor units have to be delivered over a long arc the gantry speed could be very high at the normal dose rate. If the calculated speed required exceeds the maximum speed of the gantry then it becomes necessary to reduce the dose rate to a level where this is not the case. A further requirement in rotation therapy, or in any form of dynamic therapy, is that of continuous verification. Having determined the total number of monitor units to be delivered and the length of the arc to be carried out it is possible to continuously check that the gantry is in the expected position ($\pm$ a pre-determined tolerance) throughout the exposure. This monitoring can be purely passive, in which case the rotation therapy control system is used to control the speed or dose rate and the monitoring will interrupt treatment if the tolerance is exceeded. Alternatively the monitoring can be active so that the actual position is compared with the required position and any small error, inside the tolerance, is then used to adjust the speed.

Rotation therapy need not be continuous or cover the full 360° of gantry rotation. It is often desirable to restrict the length of an arc to avoid irradiation of a critical organ such as the spinal cord. It is also sometimes necessary to change beam settings between segments, for example to insert a wedge. Note that the use of a motorized wedge during rotation therapy requires two arcs with the fraction of dose delivered in each being determined from the wedge angle and the use of dynamic wedges becomes extremely complicated.

## 16.1.2. Electron rotation therapy

Electron arc therapy has the same control requirements but somewhat different applications. The best documented application is for the irradiation of the post-mastectomy chest wall while sparing the underlying lung (McNeely *et al* 1988). The isocentre is positioned at a depth inside the chest such that the chest wall forms an approximately cylindrical surface around the gantry axis. A special electron applicator, which does not touch the patient, collimates the beam to a slit which is scanned over the surface as the gantry rotates. Final collimation is achieved by lead shielding on the skin surface. It is very important to ensure that the electron energy is chosen correctly so that the dose is not concentrated at the isocentre in the lung and that the x-ray contamination of the electron beam is as low as possible so that an unintentional x-ray rotation treatment is not delivered to the isocentre.

## 16.1.3. Conformal therapy

The term conformal therapy has been applied to techniques where particular attention is paid to 'conforming' the dose distribution to the target volume. All such techniques depend on three-dimensional identification of the target volume requiring state of the art imaging, three-dimensional treatment planning taking into account the properties of the radiation beams and the individual anatomy of each patient and in some cases assessment of the expected biological effects of the proposed treatment in addition to the dose distribution. The inclusion of biological modelling in the treatment planning process is, at the time of writing, subject to considerable uncertainty. However as the responses of different tissues are known to vary it seems logical that the introduction of biological modelling should be pursued and when the models are proved and data established it will be possible to optimize treatment plans on the basis of predicted biological effects rather than on the basis of dose distribution. Note that this is a controversial issue and many take the view that biological modelling will never reach the stage of practical application.

*16.1.3.1. Multiple fixed beams.* If non-conformal therapy is considered to be the use of rectangular beams whose dimensions are chosen to be cover the target volume then first-order conformal therapy is achieved by shaping each beam used so that it just covers the clinical target volume including an appropriate margin. The first benefit is that the reduction of the planned target volume might allow for dose escalation by exploiting the inverse relationship between tolerance dose and volume. The second benefit is that critical structures close to the clinical target volume can be shielded. In principle such shaped fields can be produced by customizing shielding blocks for each field. The alternative approach is to use an MLC (chapter

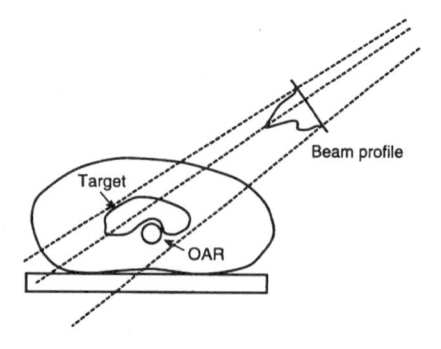

**Figure 16.1.** *A re-entrant target volume close to a sensitive organ at risk (OAR).*

6) to produce shaped fields in which case it is very easy to integrate the production of the fields with the treatment planning process. An example of computer assisted generation of MLC settings for conformal therapy has been described by Hounsell *et al* (1992). In general the beam edge created by an MLC will be stepped, with the steps corresponding to the leaf width, which is usually 1 cm, but it has been shown that the isodose surfaces at the edge of the target volume are usually considerably smoother than the geometric projection would suggest (Jordan and Williams 1994).

The use of multiple shaped fields is very effective in reducing the target volume: it can be noted that a sphere with a diameter of 8 cm formed by the intersection of multiple circular fields has a volume of 268 cm$^3$ compared with 402 cm$^3$ for the cylindrical volume formed by the rotation of an 8 cm × 8 cm field or 512 cm$^3$ for an 8 cm cube formed by the intersection of two parallel pairs of square fields.

*16.1.3.2. Multiple beams with intensity modulation.* Although the use of uniform shaped beams reduces the volume of irradiated tissue it does not necessarily result in a uniform dose distribution. In the case of co-planar beams there is a further constraint in that the boundary of the irradiated volume in the plane of the beams is always convex, so it is not possible to create re-entrant targets that might be needed to wrap a treatment volume about a sensitive organ. This problem is shown schematically in figure 16.1. In *The Physics of Three Dimensional Radiation Therapy* Webb (1993) discusses the analytical methods of determining the beam profiles which, when combined, produce concave targets. A reasonable first approximation to the required profile for each beam can be made by setting the intensity at each point to be proportional to the length of the chord projected though the target volume from that point.

As with beam shaping the production of an intensity modulated beam does not require high technology, a customized compensator (section 8.5) could be

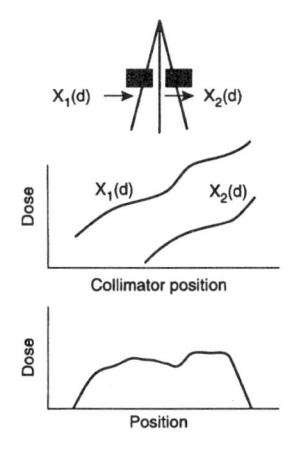

**Figure 16.2.** *A method of beam modulation by dynamic collimation: scanning slit. The separation of the two trajectories at each point in the field determines the dose at that point.*

manufactured for each beam. However if a large number of beams are used the use of compensators would be impractical. Beam intensity modulation by use of dynamic control of an MLC has been demonstrated by Yu *et al* (1995). The generation of modulated beams can be thought of as an extension of the generation of wedged fields by dynamic collimation although in this case the algorithms required to calculate the collimator trajectories are a great deal more complicated. The problem is two-dimensional and, as the trajectories for adjacent leaves will not in general be the same, the motions of all leaves must be calculated so that leaf collisions are avoided.

Several methods of modulation are being explored and it is unlikely that a single method will be optimal for all applications.

Figure 16.2 illustrates the method described by Convery and Rosenbloom (1992). Each pair of leaves generates a slit field which is scanned across the field. As the slit moves its width is varied so that points in the treatment plane are exposed for variable lengths of time and thus receive varying doses. Note that if the degree of modulation is low, the leading leaf will cross the field before the trailing leaf starts moving. In this case the term 'scanning slit' can be misleading. The principle is still that both leaves track across the beam in the same direction so that their trajectories (positions as a function of monitor units delivered) determine the dose distribution. A simpler alternative, the shrinking field method, is possible if the modulation required has a single peak. An example of use of the shrinking field method is for producing local boosts within an otherwise uniform volume and allows the boost to be delivered concurrently with the main field.

While the principle of beam intensity modulation by dynamic multileaf collimation has been established the practicalities have yet to be proved. In particular, for the scanning slit method, the accuracy of the dose delivered is critically dependent on the positional accuracy of each leaf throughout the duration of irradiation. This can be explained by considering how a uniform dose can be delivered by scanning a slit which has a constant width of, say, 2 cm across the field. If, as result of a minor mis-calibration of the leaf positioning system, the leading leaf were always 1 mm ahead of its required position and the trailing leaf were always 1 mm behind its required position then the actual width would be 2.2 cm and the dose delivered would be 10% higher than prescribed. Acceptable tolerances for machine operating conditions during beam intensity modulation are likely to be significantly tighter than for simpler techniques.

*16.1.3.3. Intensity modulated arc therapy.* Intensity modulated arc therapy or tomotherapy can be considered as the inverse of CT scanning; however the inverse analogy is not complete. The convolution kernels required to be applied to the diagnostic x-ray transmission profiles in order to remove the $1/r$ blurring function include negative values and thus back-projection to form a faithful image of the object scanned includes the projection of negative as well as positive numbers. In theory any arbitrary dose distribution could be generated by projecting therapy beams with profiles calculated from the line integrals of dose deposited along each ray convolved with same functions used for CT. Unfortunately although the manipulation of negative numbers in a computer is easy, in the case of a therapy beam projection the beam modulation is constrained by allowing only positive intensities. However if it is possible to accept limitations on the dose distribution, specifically to accept some dose at all points outside the target volume, it is possible to calculate sets of profiles which will produce quite complex dose distributions within a patient. Such calculations can either be analytical (Brahme 1988) or iterative (Webb 1991). While conceptually tomotherapy is appealing its application is not straightforward as it is necessary to modulate the beam during rotation of the gantry.

One approach to this problem has been developed by the Nomos company (Pittsburgh, PA, USA) and makes use of a multivane intensity modulation system which consists of forty tungsten leaves arranged in two rows of twenty. Each leaf is pneumatically powered and can occlude a 1 cm × 1 cm area of the beam, whose maximum size is 20 cm × 2 cm. A rotation treatment is split into many segments of a few degrees each and the leaves are opened for a predetermined fraction of the time taken to irradiate each segment. The limitation is that the three-dimensional dose distribution has to be built up from a series of contiguous 2 cm wide 'inverse CT' slices.

An alternative approach is to calculate the required two-dimensional intensity distributions as a function of gantry angle and then to decompose the resulting three-dimensional matrix (intensity as a function of field $-x$, field $y$ and gantry $-\theta$) into a set of shaped static fields for each of which a simple rotation treatment is delivered in turn. Integration of the segments will then result in the dose distribution required. This approach has been described by Yu (1995).

The difference between delivering a large number of multiple beams each with intensity modulation and delivering a large number of rotations using uniform shaped beams is likely to be very small but the practicalities including considerations of speed, ease of verification and safety might prove decisive.

## 16.2. STEREOTACTIC RADIOSURGERY

The term radiosurgery has been used to describe radiation therapy where the doses are so high that all tissues are ablated. This is different from radiotherapy, where the aim is to produce more subtle effects with the intention of killing tumour cells while sparing most of the normal cells, even those within the target volume. Radiosurgical ablation has been shown to be effective in treating otherwise inoperable lesions within the brain, in particular arteriovenous malformations (AVMs). A dose of between 15 and 20 Gy delivered as a single fraction is sufficient to block the damaged blood vessels and avoid the risk of further haemorrhage. Clearly at this dose level the dose distribution has to be very tightly confined to the target tissues in order to minimize morbidity to normal brain tissue.

AVMs are typically between 1 and 3 cm in diameter and it is desirable to reduce the dose to below brain tolerance within 1 cm of the target volume. The prerequisites for radiosurgery are therefore precise methods of localization of the lesion within the brain, precise methods of relocation of the lesion in the frame of reference of the therapy machine and very precise methods of collimation and beam delivery. Perversely, as long as these geometric requirements are met the precision in actual dose delivered is less critical than is the case in conventional radiotherapy.

Precision in localization is achieved by the use of a stereotactic frame for both the imaging procedures during which the site of the lesion is determined and the irradiation procedure during which the frame is used for patient fixation. Stereotactic frames are commonly used in neurosurgery to allow surgeons very precise control of their surgical instruments. In this case a modified rigid open frame, which carries radio-opaque fiduciary markers, is fixed to the skull by screws, a procedure that can be carried out under local anaesthetic. Localization of the lesion relative to the stereotactic frame is carried out by CT scanning the patient in the frame. In most cases the frame is not removed from the patient until after the radiosurgery is completed.

A very specialized radiosurgery unit, the gamma knife, using multiple radioactive sources was developed for this type of treatment by Leksell (1983) and subsequently many linear accelerators have been adapted to

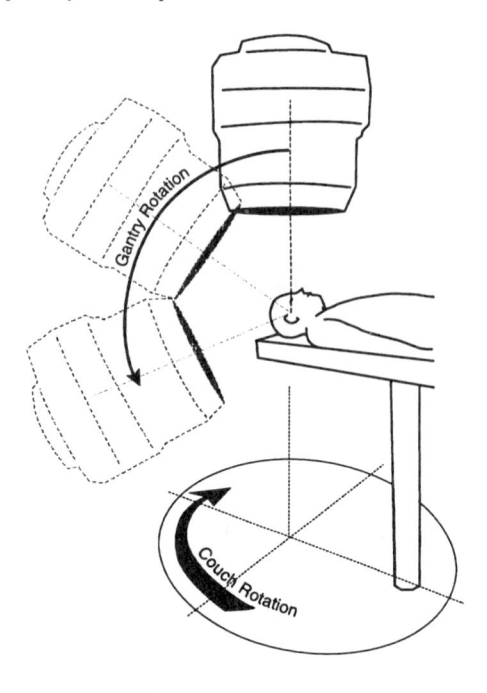

**Figure 16.3.** *The use of multiple non-co-planar arcs for delivering stereotactic radiosurgery.*

perform these treatments (Friedman and Bova 1989, Lutz *et al* 1988). The first requirement is to link the frame of reference in which the location of the tumour is known (the stereotactic frame) to the frame of reference of the accelerator. If the isocentric accuracy of the normal treatment couch is acceptable then all that is needed is attachment of the stereotactic frame to the couch in the same way as other fixation devices are used. If the treatment couch is not sufficiently precise then alternative mounting must be provided. Collimation is usually provided by a supplementary circular collimator mounted so that it rotates very accurately around the isocentre. Secondary collimators are needed to ensure precision, a minimum penumbra and to collimate the beam to much smaller sizes than are usual in normal radiotherapy. Graham *et al* (1991) studied the different irradiation techniques for delivering radiosurgery and concluded that for spherical lesions the best sparing of normal tissues is achieved by the use of multiple non-co-planar arcs as illustrated in figure 16.3. The alternative for non-spherical lesions is to use multiple shaped non-co-planar static beams.

Clearly the use of such a technique puts high demands on the quality assurance of the geometric aspects of the linear accelerator. It is a very time consuming procedure but is nevertheless being explored for wider application particularly in the irradiation of solitary metastatic deposits in which case irradiation is to conventional radiotherapy doses and relocatable stereotactic frames are being used so that fractionation of treatment is possible.

# CHAPTER 17

# CONCLUSION

Radiation therapy has been a useful method of treating cancer for more than sixty years. Its development has been dependent on the availability of suitable radiation sources and the means of measuring radiation dose. It is perhaps surprising that after such a long period the subject is still in a state of growth, in terms of the number of treatment units being installed in the developed and developing countries, in research and development in the techniques for producing and measuring the required radiations and in adapting the radiation dose pattern to the needs of the individual patient.

In the first phase of development, radiotherapists were very limited by the penetrating properties of the available x-ray beams and one can only admire what was achieved with the equipment that was available up to the early 1950s (Paterson 1956). At this time linear accelerators, betatrons and cobalt-60 beam units started to become available to radiotherapy departments and in the following decades were installed in substantial numbers round the world. Most people would now agree that the competition between these three types of radiation source has been won by linear accelerator technology, and that both the numbers and types in use continue to grow rapidly.

Other types of electron accelerator such as the microtron (Svensson *et al* 1977), and its higher-energy derivative the racetrack microtron (Rosander *et al* 1982), have come into clinical use though the basic design of this machine was proposed by Vexsler as long ago as 1944. Their development as a generally available clinical tool in competition with the linear accelerator is not yet foreseeable.

The development of linear accelerator technology has been paralleled by changes in the practice of radiotherapy, that is, in the methods by which the radiation is applied to patients. These are described in clinical radiotherapy text books, for example that by Pointon (1990).

The present book is aimed at those responsible for the operation of linear accelerators in hospitals, mainly hospital physicists and engineers, and is intended to outline the basic technology and the main design decisions which determine the properties of the currently available machines.

# Conclusion

It should be apparent from earlier chapters that future developments in linear accelerators will be mainly in the application of computers to the control and safety systems rather than the beam production systems.

Conformal therapy has developed considerably in the last ten years and it is likely that the tools for the simpler applications will become standard parts of the next generation of accelerators rather than optional extras. However it should be noted that it has taken twenty years for the ideas, which were the basis of the early development of work by Green (1965) and Davy *et al* (1975), to bear fruit.

Modern computer technology will enable developments to take place on a much shorter time scale. It will therefore be very important to test the scientific and engineering principles and then the clinical efficacy of each new development as it arises. Failure to do so could result in better linear accelerators rather than linear accelerators capable of providing better treatments.

If we are right about the speed of development, the new machines being brought into clinical service in ten years time will be based on an idea which has not yet been thought of. Do you know what it is?

# REFERENCES

American Association of Physicists in Medicine (AAPM), 1983 AAPM Task Group 21: a protocol for the determination of absorbed dose from high energy photon and electron beams *Med. Phys.* **10** 741–71
—— 1994 AAPM Task Group 39: The calibration and use of plane-parallel ionization chambers for the dosimetry of electron beams *Med. Phys.* **21** 1251–60
Arnot R N, Willets R J, Batten J R and Orr J S 1984 X-ray image intensifier and TV camera for imaging transverse sections in humans *Br. J. Radiol.* **57** 47–55
Associazione Italiana di Fisica Biomedica (AIFB) 1988 Protocollo per la dosimetria di base nella radioterapia con fasci di fotoni ed elettroni con Emax fra 1 e 40 MeV *Fis. Biomed.* 6
Berger M J and Seltzer 1964 Studies in penetration of charged particles in matter *National Academy of Sciences Publication* 1133
Boag J 1966 *Radiation Dosimetry* vol 2, ed F H Attix and W C Roesch (New York: Academic)
Botman J I M, Bates T and Hagedoorn H L 1985 A double focusing magnet system for a medical linear electron accelerator *Nucl. Instrum. Methods Phys. Res.* B **10/11** 796–8
Brahme A 1988 Optimisation of stationary and moving beam radiation therapy techniques *Radiother. Oncol.* **12** 129–40
British Journal of Radiology 1983 Central axis depth dose data for use in radiotherapy *Supplement* 17
—— 1989 Treatment simulators *Supplement* 23
—— 1996 Central axis depth dose data for use in radiotherapy *Supplement* 25
British Standards Institute 1984 Specification for medical linear accelerators in the range 1 MeV to 50 MeV BS 5724 (superseded by IEC 601-2-1 (1995))
Carpenter L G 1983 *Vacuum Technology* 2nd edn (Bristol: Hilger)
Comite Francais Mesure des Rayonnements Ionisants (CFMRI) 1987 Recommendations pour la mesure de la dose absorbée en radiotherapie

dans les faisceaux de photons et d'electrons d'energie comprise entre 1 MeV et 50 MeV *CFMRI Report* 2

Connor F R 1982 *Wave Transmission* (London: Arnold)

Convery D and Rosenbloom M 1992 The generation of intensity modulated beams for conformal therapy by dynamic multi leaf collimation *Phys. Med. Biol.* **37** 1359–74

Cunningham J R, Shrivastava P N and Wilkinson J M 1972 Program irreg-calculation of dose from irregularly shaped radiation beams *Comput. Prog. Biomed.* **2** 192–9

Davy T J, Johnson P H, Redford R and Williams J R 1975 Conformation therapy using a tracking cobalt unit *Br. J. Radiogr.* **48** 122–30

Elliot J H, 1972 *Handbook of Microwave Techniques and Equipment* (New York: Prentice-Hall)

Enge H E 1963 Achromatic magnet mirror for ion beams *Rev. Sci. Instrum* **34** 385–9

Epp E R, Laughlin J S, Swanson W P and Bond V P 1984 Neutrons from high energy x-ray medical accelerators: an estimate of risk to the radiotherapy patient *Med. Phys.* **11** 231–41

Evans P M, Hansen V N, Swindell W, Torr M, Mayles W P M, Neal A J, Brown S, and Yarnold J R 1994 The use of portal imaging to design tissue compensators for radiotherapy of the breast *Proc. 11th Int. Conf. on the Use of Computers in Radiation Therapy* ed A R Hounsell, J M Wilkinson and P C Williams pp 118–9

Friedman W A and Bova F J 1989 The University of Florida radiosurgery system *Surg. Neurol.* **32** 334–42

Fry D W, Harvie R B S, Mullett L B and Walkinshaw W W 1947 Travelling wave linear accelerator for electrons *Nature* **160** 35

Galbraith D M, Rawlinson J A and Munro P 1984 Dose errors due to charge storage in electron irradiated plastic phantoms *Med. Phys.* **11** 253

Graham J D, Nahum A E and Brada M 1991 A comparison of techniques for stereotactic radiotherapy by linear accelerator based on 3-dimensional dose distributions *Radiother. Oncol.* **22** 29–35

Green A 1965 Tracking Cobalt Project *Nature* **207** 1311

Greene D 1983 The cost of radiotherapy treatments on a linear accelerator *Br. J. Radiol.* **56** 189–91

Greene D, Chu G and Thomas D W 1983 Dose levels outside radiotherapy beams *Br. J. Radiol.* **56** 543–50

Greene D and Fallas P 1985 Long term performance of linear accelerators *Br. J. Radiol.* **58** 556

Greene D and Massey J B 1961 Some measurements of the absorption of 4 MV x-rays in concrete *Br. J. Radiol.* **34** 389

Greening J R 1985 *Fundamentals of Radiation Dosimetry* 2nd edn (Bristol: Hilger)

*Handbook of Radiological Protection Part I Data* 1971 (London: HMSO)

# References

Health and Safety Commission 1985 *Approved Code of Practice for the Protection of Persons against Radiation arising from any Work Activity* (London: HMSO)

Health and Safety Executive 1992 Fitness of equipment used for medical exposure to ionising radiations *HSE Guidance Note* PM77 (London: HMSO)

Holloway A F and Cormack D V 1980 Radioactive and toxic electricity production by a medical linear accelerator *Health Phys.* **38** 673–7

Hospital Physicists' Association (HPA) 1971 A practical guide to electron dosimetry (5–35 MeV) *HPA Report* 4

—— 1975 A practical guide to electron dosimetry for radiotherapy purposes *HPA Report* 13

—— 1983 Revised code of practice for the dosimetry of 2–35 MV x-rays and caesium-137 and cobalt-60 gamma ray beams *Phys. Med. Biol.* **28** 1097–104

Hounsell A R, Sharrock P J, Moore C J, Shaw A J, Wilkinson J M and Williams P C 1992 Computer assisted generation of multileaf collimator settings for conformation therapy *Br. J. Radiol.* **65** 321–6

Hounsell A R and Wilkinson J M 1990 Tissue standard ratios for irregularly shaped radiotherapy fields *Br. J. Radiol.* **63** 629–34

Howard-Flanders P and Newbery G R 1950 A gantry type of mounting for high voltage x-ray therapy equipment *Br. J. Radiol.* **23** 355–7

Hubbell J H and Berger M J 1968 *Engineering Compendium on Radiation Shielding* (Berlin: Springer)

Institute of Physical Scientists in Medicine (IPSM) 1988 Commissioning and quality assurance of linear accelerators *IPSM Report* 54

—— 1990 Code of practice for high energy photon therapy dosimetry based on the NPL absorbed dose calibration service *Phys. Med. Biol.* **35** 1355–60

Institute of Physics and Engineering in Medicine (IPEM) 1997 Design of radiotherapy treatment room facilities *IPEM Report* 75

Institution of Physics and Engineering in Medicine and Biology (IPEMB) 1996 The IPEMB code of practice for electron dosimetry for radiotherapy beams of initial energy from 2 to 50 MeV based on an air kerma calibration *Phys. Med. Biol.* **41** 2557–603

International Atomic Energy Agency (IAEA) 1987 Absorbed dose determination in photon and electron beams *IAEA Technical Report* 277

International Commision on Radiation Units and Measurements (ICRU) 1969 Radiation dosimetry: x-rays and gamma rays with maximum photon energies between 0.6 and 50 MeV *ICRU Report* 14

—— 1972 Radiation dosimetry: electrons with initial energies between 1 and 50 MeV *ICRU Report* 21

—— 1976 Determination of absorbed dose in a patient irradiated by x- or gamma rays in radiotherapy procedures *ICRU Report* 24

—— 1982 The dosimetry of pulsed radiation *ICRU Report* 34

International Commission on Radiological Protection (ICRP) 1977 *Annals of the ICRP* vol 1 *(ICRP Publication 26)* (Oxford: Pergamon)

—— 1983 Protection against ionising radiation from external sources used in medicine *Annals of the ICRP* vol 39 *(ICRP Report 33)* (Oxford: Pergamon)

—— 1991 *1990 Annals of the ICRP: Recommendations of the International Commission on Radiological Protection (ICRP Publication 60)* (Oxford: Pergamon)

International Electrotechnical Commission (IEC) 1981 Specification for safety of medical electron accelerators in the range 1–50 MeV *IEC Publication* 601-2-1

—— 1989a Methods of declaring functional performance characteristics of medical electron accelerators in the range 1 MeV to 50 MeV IEC976

—— 1989b Methods of declaring functional performance characteristics of medical electron accelerators in the range 1 MeV to 50 MeV Supplement 1. Guide to functional performance values IEC977

*Ionising Radiations Regulations* 1985 (London: HMSO)

Johns H E and Cunningham J R 1983 *The Physics of Radiology* (Sprinfield, IL: Thomas)

Jordan T J and Williams P C 1994 The design and preformance characteristics of a multileaf collimator *Phys. Med. Biol.* **39** 231–51

Karzmark C J 1984 Advances in linear accelerator design for radiotherapy *Med. Phys.* **11** 105–28

Karzmark C J, Nunan C S and Tanabe E 1993 *Medical Linear Accelerators* (Springfield, IL: McGraw-Hill)

Karzmark C J and Pering N C 1973 Electron linear accelerators for radiation therapy: history principles and contemporary developments. *Phys. Med. Biol.* **18** 321

Kersey R W 1979 Estimation of neutron and gamma radiation doses on the entrance mazes of SL75 20 linear accelerator treatment rooms *Med. Mundi* **24** 151

Klevenhagen S C 1983 *The Physics of Electron Beam Therapy* (Bristol: Institute of Physics)

Leksell L 1983 Stereotactic radiosurgery *J. Neurol. Neurosurg Psychiatry* **46** 797–803

Lillicrap S C, Owen B, Williams J R and Williams P C 1990 Code of practice for high energy photon therapy dosimetry based on the NPL absorbed dose calibration service *Phys. Med. Biol.* **35** 1355–60

Lundberg D A 1971 *Br. J. Radiol.* **44** 708

Lutz W, Winston K R and Maleki N 1988 A system for stereotactic radiosurgery with a linear accelerator *Int. J. Radiat. Oncol. Biol. Phys.* **14** 373–81

McCall R C 1981 Neutron sources and their characteristics *4th Symp. on Neutron Dosimetry in Biology and Medicine* Euratom

McGinley P H, Wood M, Mills M and Rodriguez R 1976 Dose levels due to neutrons in the vicinity of high energy accelerators *Med. Phys.* **3** 397

McNeely L K, Jacobson G M, Leavitt D D and Stewart J R 1988 Electron arc therapy: chest wall irradiation of breast cancer patients *Int. J. Radiat. Oncol. Biol. Phys.* **14** 1287–94

Meredith W J and Massey J B 1972 *Fundamental Physics of Radiology* (Bristol: Wright)

Miller C W 1954 An 8 MeV linear accelerator for x-ray therapy *Proc. IEE* **101** 207–22

Mooijweer H 1971 *Microwave Techniques* (London: Macmillan)

Morton E J, Swindell W, Lewis D G and Evans P M 1991 A linear array scintillation crystal photo diode detector for megavoltage imaging *Med. Phys.* **18** 681–91

Mould R F 1985 *Radiotherapy Treatment Planning* 2nd edn (Bristol: Hilger)

National Council on Radiation Protection and Measurements (NCRP) 1976 Structural shielding design and evaluation for medical use of x-rays and gamma rays up to 10 MeV *Report 49*

—— 1977 Radiation protection design guidelines for 0.1–100 MeV particle accelerator facilities *NCRP Report 51*

—— 1984 Neutron contamination from medical electron accelerators *NCRP Report 79*

National Radiological Protection Board 1988 Guidance notes for the protection of persons against ionising radiations arising from medical and dental use.

Nederlandse Commissie voor Stralingsdosimetrie (NCS) 1986 Code of practice for dosimetry of high energy photon beams *NCS Report 2*

—— 1989 Code of practice for dosimetry of high energy electron beams *NCS Report 5*

Nordic Association of Clinical Physics (NACP) 1980 Procedures in external radiation therapy dosimetry with electron and photon beams with maximum energies betweeen 1 and 50 MeV *Acta Radiol. Oncol.* **19** 55–79

—— 1981 Supplement to the recommendations of NACP 1980: electron beams with mean energies at the phantom surface below 15 MeV *Acta Radiol. Oncol.* **20** 401–15

Paterson R 1956 *The Treatment of Malignant Disease by Radium and X-rays* (London: Arnold)

Podgorsak E B, Rawlinson J A and Johns H E 1975 X-ray depth doses from linear accelerators in the energy range from 10 to 32 MeV *Am. J. Roentgenol.* **123** 182–91

Pointon R C S (ed) 1990 *The Radiotherapy of Malignant Disease* (Berlin: Springer)

Redpath A T, Williams J R and Thwaites D I 1993 *Radiotherapy Physics* (Oxford: Oxford Medical Publications) ch 8

Rogers D W O and Van Dyke G 1981 Use of a neutron remmeter to measure leakage neutrons from medical linear accelerators *Med. Phys.* **8** 163–6

Rosander S, Sedlacek M and Wernholm O 1982 The 50 MeV racetrack microtron at the Royal Institute of Technology, Stockholm *Nucl. Instrum. Methods* **204** 1–20

Rosenbloom M E, Killick L J and Bentley R E 1977 Verification and recording of radiotherapy treatments using a small computer *Br. J. Radiol.* **50** 637–44

Sociedad Española de Fisica Medica (SEFM) 1987 Procedimientos recomendados para la dosimetria de fotones y electrones de energias compendidas entre 1 MeV y 50 MeV en radioterapia de haces extraños *SEFM Report 2*

Sloan D H and Lawrence E O 1931 The production of heavy high speed ions without the use of high voltages *Phys. Rev.* **38** 2021

Stewart J G and Jackson A W 1975 The steepness of the dose response curve both for tumour cure and normal tissue injury *Laryngoscope* **85** 1107–11

Svensson H, Jonsson L, Larsson G, Brahme A, Lindberg B and Reistad D 1977 A 22 MeV microtron for radiation therapy *Acta Radiol. Ther.* **16** 145

Tanabe E and Hamm R 1985 Compact multi energy electron linear accelerator *Nucl. Instrum. Methods* B **10/11** 871–6

Terman F E 1943 *Radio Engineers Handbook* (New York: McGraw-Hill)

van Herk M and Meertens H 1988 A matrix ionisation chamber imaging device for on line patient setup verification during radiotherapy *Radiother. Oncol.* **11** 369–78

Vexsler V I 1944 *Proc USSR Acad. Sci.* **43** 346

Webb S 1988 *The Physics of Medical Imaging* (Bristol: Institute of Physics)

—— 1991 Optimisation by simulated annealing of three dimensional conformal treatment planning for radiation fields defined by a multileaf collimator *Phys. Med. Biol.* **36** 1201–26

—— 1993 *The Physics of Three Dimensional Radiation Therapy* (Bristol: Institute of Physics)

Wideroe R 1928 Uber ein neues prinzip zur herstellung hoher spannungen *Arch. Elektrotech.* **21** 387

Williams J R and Thwaites D I 1993 *Radiotherapy Physics* (Oxford: Oxford Medical Publications)

Williams P C, Greene D, Burns J E and Lillicrap S C 1983 A code of practice for the dosimetry of high energy x-rays *Phys. Med. Biol.* **28** 1097–104

World Health Organisation (WHO) 1988 *Quality Assurance in Radiotherapy* (Geneva: WHO)

Yu C X 1995 Intensity modulated arc therapy with dynamic multileaf collimation: an alternative to tomotherapy *Phys. Med. Biol.* **40** 1435–50

Yu C X, Symons M J, Du M N, Martinez A A and Wong J W 1995 A method for implementing dynamic photon beam intensity modulation using independent jaws and a multileaf collimator *Phys. Med. Biol.* **40** 769–88

# INDEX

Accelerator
  bunches section, 16, 18
  construction of waveguide, 24
  definition, 1
  historical background, 10
  standing wave, 17
  travelling wave, 13
Automatic frequency control (AFC), 38, 42
Axis lights, 115, 211

Beam bending magnet
  90°, 66
  112.5°, 68
  270°, 67
Beam defining collimators
  independent movement, 79
  multileaf, 80
  symmetrical movement, 79
Beam flattening filter, 75
Beam focusing coils, 63
Beam monitor, 94
Beam shaping
  electron fields, 119
  x-ray fields, 116
Beam stabilization, 98
Beam steering coils, 98
Beam transport, 62
Bunches, 16

Calibration
  monitor, 186
  phantom, 179, 215
  record keeping, 223
  results, 219
Cavities
  tuned, 38
Circuits
  to gantry, 126
  to treatment head, 92
Collimator
  beam defining
    complex shaped fields, 80
    electron field, 88
    sample shaped fields, 79
  primary, 78
Computer control, 152
Computed tomography, 241, 245
Control circuit, 148
Control consoles, 160, 209
Control systems
  automatic, 146
  automatic select and confirm, 146
  select and confirm, 145
Cooling systems
  basic function, 5
  details, 52
  servicing, 224
Costs
  operating, 232

$D_{10}$ specification for x-rays, 183, 185
$d_{80}$ specific for electrons, 183, 186

Data
  accelerator performance, 228
  attenuation, electron, 119
  attenuation, x-ray, 77
  attenuation in concrete, 205
Door knob transformer, 17
Dose distribution
  across an x-ray field, 76, 108
  with depth, 182
Dose gradient of wedge filters, 85
Dose monitoring, 101
  specification, 181
Dose rate monitor, 104
Dose rate specification, 189
Dosimeter calibration, 186
Dosimetry, 167
Dosimetry units, definition, 23
Dynamic wedge, 87

Electroforming, 24
Electron beam
  calibration, 189, 217
  collimator, 88
  current, 23
  quality, 183
  scanned, 90
Electron bunches, 47
Electron gun
  diode, 58
  triode, 58
Electron gun heater supply
  diode, 59
  triode, 61
Emergency, 211
Equivalent dose definition, 202

Field flatness
  checks, 221
  specification, 185
Field size
  definition electron, 111
  definition x-rays, 108
Filter
  beam flattening, 75
  wedge, 83
Focusing coils, 63
Foil, electron scattering, 88, 90
Four-port circulates, 37
Four-port couples, 37
Frequency
  control (AFC), 38, 42
  monitor, 39, 42
  operating, 1, 29
  pulse repetition (PRF), 45
  selection, 39, 40

Gantry
  drum mounting, 123
  movement control, 125
  pendulum mounting, 126
  speed control, 135
Gray, definition, 23

Hydrogen thyratron, 43

Interlock
  circuit, 148
  machine interlock list, 141
Ion pump, 49
Isocentre, 117
Isocentric mounting, 121
Isodose charts, 183
Isolator, 35

Klystron, 33
  frequency control, 42

Machine control, parameter lists, 141, 165
Magnet
  beam bending, 7
  90° system, 66
  112.5°, 68
  270°, 67
Magnetron, 30
  cooling, 5
  frequency control, 38
  heater supply, 46
Maze design, 201

Mechanical pointers, 112, 175
Mechanical tests
 gantry, 177
 patient support systems, 177
 treatment head, 175
Microwave generator
 klystron, 33
 magnetron, 30
Modulator, 43
 safety, 225
 servicing, 225
Monitor, 94
Movement control
 couch, 135
 whole system, 135
MR scanning, 245
Multipass accelerator, 21

Neutron production, 198
 neutron door, 209
 neutron shielding, 208

Optical pointers
 back pointer, 116
 beam centring, 115
 SSD scale, 116

Patient couch, 131
Patient prescription, 142
Patient safety, 136
Patient support system
 basic description, 8
 couch, 128
 limited movement, 131
 maximum movement, 129
 novel system, 132
 testing of, 177
Patient viewing, 211
Phase control, 40
Phase shifter, 36, 40
Portal imaging
 detector array, 172
 film, 169
 fluorescent screen, 170

Pretzel magnet, 68
Primary collimator, 78
Pulse forming network (PFN), 43
Pulse repetition frequency (PRF)
 generator, 45
Pulse transformer
 parallel windings, 46
Pulse voltage control, 45

$Q$-spoiling, 45

Radiation control, 94
Radiation output measurement, 178, 215
Radiation outside useful beams, 194
Radiation protection, 204
Radioactivity production, 197
Record keeping
 dosimeter calibrations, 223, 226
 machine operation, 226
Relation, x-ray field to optical field, 108
RF loads, 37
Rotary joints in waveguide, 38, 126

Scales
 on mechanical systems, 136
 on treatment head, 67
 SSD 114
Servicing
 computer systems, 226
 cooling systems, 224
 electronic, 225
 mechanical, 224
Shadow blocks, moulded, 118
Shadow tray, 116
Simulator, 235
Slalom system, 68
Spares, 227
Standing wave accelerator, 17
Steering coils, 62

Target, 75
'Tennis racket' 132
Thyratron, 43

Transmission waveguide, 12, 13
Transverse axial tomograph, 242
Treatment controls, 141
Treatment room design, 199
  access, 201, 212
  control of access, 210
  lighting, 211
  maze design, 201
  radiation protection, 204
  shielding data, 205
Tuned cavities for frequency control, 38

Vacuum system, 48
  servicing, 225
  ventilation, 211
  verification, 164

Warning lights, 210
Water cooling, 52
Waveforms
  magnetron, 47
  on PFN, 44
  radiation output, 47
Waveguide
  construction, 24
  side coupled, 16
  theory, 11
  transmission, 12, 34
  window, 35
Wedge angle, definition, 83
Wedge factors, 213, 222
Wedge filter, 83
  dynamic, 87
  motorized, 85
Window
  waveguide, 35

X-ray, attenuation coefficients, 77
X-ray, quality, 183, 185, 211
X-ray calibration, 187
X-ray target, 75